Studies in Computational Intelligence

Volume 947

Series Editor

Janusz Kacprzyk, Polish Academy of Sciences, Warsaw, Poland

The series "Studies in Computational Intelligence" (SCI) publishes new developments and advances in the various areas of computational intelligence—quickly and with a high quality. The intent is to cover the theory, applications, and design methods of computational intelligence, as embedded in the fields of engineering, computer science, physics and life sciences, as well as the methodologies behind them. The series contains monographs, lecture notes and edited volumes in computational intelligence spanning the areas of neural networks, connectionist systems, genetic algorithms, evolutionary computation, artificial intelligence, cellular automata, self-organizing systems, soft computing, fuzzy systems, and hybrid intelligent systems. Of particular value to both the contributors and the readership are the short publication timeframe and the world-wide distribution, which enable both wide and rapid dissemination of research output.

Indexed by SCOPUS, DBLP, WTI Frankfurt eG, zbMATH, SCImago.

All books published in the series are submitted for consideration in Web of Science.

More information about this series at http://www.springer.com/series/7092

Stefka Fidanova

Ant Colony Optimization and Applications

Springer

Stefka Fidanova
Institute of Information and Communication
Technology
Bulgarian Academy of Sciences
Sofia, Bulgaria

ISSN 1860-949X ISSN 1860-9503 (electronic)
Studies in Computational Intelligence
ISBN 978-3-030-67379-6 ISBN 978-3-030-67380-2 (eBook)
https://doi.org/10.1007/978-3-030-67380-2

This Springer imprint is published by the registered company Springer Nature Switzerland AG
The registered company address is: Gewerbestrasse 11, 6330 Cham, Switzerland

Foreword

Combinatorial optimization problems are extremely difficult from a computational point of view. There is no universal method that can solve them in efficient way. Even different variants or instances of the same problem may be solved by different methods. Usually the efficient algorithms for solving combinatorial optimization problems are problem dependent.

In this book, Stefka Fidanova focuses on Ant Colony Optimization (ACO), among the best approaches for solving combinatorial optimization problems. ACO differs from other methods in that it is a constructive method and has obtained distinguished results on some applications.

The book is the result of the authors work, over the last 20 years, on applying ACO to a variety of problems coming from real life and industry. The author introduces the ACO algorithm and its variety in the first and second chapter. Some of the variants like ACO with additional reinforcement and start strategies are proposed by the author.

Following chapters are devoted to the application of the ACO on a variety of problems. I will mention Multiple Knapsack Problem, Grid Scheduling Problem, GPS Surveying Problem, E. coli Cultivation Modeling, Wireless Sensor Network Positioning, Image Edges Detection, and Workforce Planning. All of them are very complex combinatorial optimization problems.

Stefka Fidanova has written a book, which targets students, researchers, and practitioners. The students and the researchers will acquire knowledge about ACO and its possible applications as well as practitioners will find new ideas and solutions of their combinatorial optimization problems. This book is well written and full of new ideas. It will provoke our curiosity to seek and learn even more.

2020

Prof. Krassimir Atanassov
Bulgarian Academy of Sciences
Sofia, Bulgaria

Foreword

Preface

Collective intelligence allows animals to overcome difficulties and obstacles in nature, even when they do not have a high level of individual intelligence. Bees, ant colonies, bird , fish passages, etc. can be given as examples of group intelligence. This phenomenon has inspired scientists to imitate nature in their quest to solve complex problems coming from real life and industry.

The unique behavior of ants in nature and their ability to always find the shortest path between the nest and the food source, gives the idea for the creation of the ants method. This book is a result of twenty years work of the author on improvement of ant colony optimization method and its application on various real life and industrial problems.

The book includes the structure of the ant colony optimization and author's involvement on it. The application of ant colony optimization include knapsack problem, grid scheduling problem, GPS surveying problem, bioreactor modeling problem, wireless sensor network positioning, image edges detection, and workforce planning.

The book can be useful for both researchers and practitioners.

Sofia, Bulgaria
2020

Stefka Fidanova

Acknowledgements

I wish to convey my most sincere gratitude to all my colleagues from Institute of Information and Communication Technologies at Bulgarian Academy of Science and Krassimir Atanassov, Olimpia Roeva and Vassia Atanassova from Institute of Biophysics and Biomedical Engineering at Bulgarian Academy of Sciences.

I want to thank the foreign scientists I have worked with over the years, Enrique Alba, Gabriel Luque (Spain), Marco Dorigo (Belgium), Antonio Mucherino (France), Marcin Paprzycki, Maria Ganzha (Poland), and Petrica Pop (Romania).

Special thanks are due to Janusz Kacprzyk (Poland), the editor-in-chief of this book series for kind consideration and help in arranging and running this book publication project.

And last but not least, I would like to thank my family and especially my husband who has been encouraging and stimulating all my scientific research.

This work has been accomplished with the financial support by the Grant No BG05M2OP001-1.001-0003, financed by the Science and Education for Smart Growth Operational Program (2014–2020) and co-financed by the European Union through the European structural and Investment funds and by the Bulgarian Scientfic Fund by the grant DN 12/5.

Sofia
2020

Stefka Fidanova

Contents

Chapter 1
Introduction

We can learn a lot by observing nature. There is no waste with it. Everything is done in the most economical, optimal way. Particularly impressive is the collective intelligence of a group of individuals working together. Bees, ant colonies, bird flocks, fish passages, etc. can be given as examples of group intelligence. Animals that do not have a high level of individual intelligence deal with difficult problems using a collective approach. This gaves scientists the idea to create algorithms inspired by nature, mimicking the collective intelligence of some animals. These are the so-called metaheuristic methods.

There are complex optimization problems coming from real life and industry that require large computing resources to find close to the optimal solution. For the most part, these are combinatorial optimization problems. Exact methods and traditional numerical methods are not suitable for this type of problems. In this case, the only possibility is metaheuristic methods. Their advantage is quickly finding a good solution. Their disadvantage is that their accuracy is not guaranteed. In cases where the accuracy of the solution can be compromised and it is more important to find it quickly, metaheuristic methods are preferable.

The unique behavior of ants in nature and their ability to always find the shortest path between the nest and the food source, gaves the idea for the creation of the ants method [1–4]. The author of this book is pleased to work with Marco Dorigo, the inventor of ant colony optimization method, and his team. This collaboration led to the application of the method on a variety of problems, showing its wide possibilities. The author has used the apparatus of the generalized nets for modeling Ant Colony Optimization and it gives him an ideas for improvement of the algorithm and its applications [5].

This book contains various applications of the ant method that the author has worked on. It is a summary of many years of work. The structure of the book is as follows. The book consists of Introduction and 8 chapters.

Chapter 2 describes the structure of the ant method, its main characteristics and parameters. The different variants of the algorithm are given there.

S. Fidanova, *Ant Colony Optimization and Applications*, Studies in Computational Intelligence 947, https://doi.org/10.1007/978-3-030-67380-2_1

Chapter 3 is devoted to the Knapsack Problem and its solving with ant method. Different approaches are compared. Various options are proposed for compiling the heuristic model. A variety of ways to calculate the probability of selection are compared.

A current task in recent years is Grid scheduling problem. An algorithm based on the ant method for solving it is proposed in Chapter 4. The algorithm can work in dynamic way.

It's hard to imagine a temporary life without the use of GPS. Chapter 5 offers several options for optimizing with the ant method to solve the GPS surveying problem. An intercriteria analysis of the behavior of the algorithm is made.

One of the most successful ways to produce medical substances is to use microorganisms. Chapter 6 uses the ant method to model a bioreactor. Several variants of the algorithm are proposed, as well as its combination with other heuristic algorithms.

The widespread use of wireless communications requires the creation of wireless networks with an optimal number of nodes and using as little energy as possible. This is a problem that requires optimization by two criteria. Chapter 7 offers several variants of algorithms based on the ant method for solving it.

Image edges detection is not a typical optimization problem. To solve it in Chapter 8, an algorithm was developed using ideas from the ant method. Parameters have been added with which one can set how detailed the found edges should be.

Workforce planning is a difficult optimization task with many limitations. In Chapter 9, an algorithm is proposed based on the ant method for solving it. The influence of the algorithm parameters is studied.

References

1. Bonabeau, E., Dorigo, M., Theraulaz, G.: Swarm Intelligence: From Natural to Artificial Systems. Oxford University Press, New York (1999)
2. Dorigo, M., Di Caro, G.: The ant colony optimization meta-heuristic. In: Corne, D., Dorigo, M., Glover, F. (eds.) New Idea in Optimization, pp. 11–32. McGrow-Hill, London (1999)
3. Dorigo, M., Gambardella, L.M.: Ant colony system: a cooperative learning approach to the traveling salesman problem. IEEE Trans. Evol. Comput. **1**, 53–66 (1997)
4. Dorigo, M., Stutzle, S.: Ant Colony Optimization. MIT Press, Cambridge (2004)
5. Fidanova, S., Atanassov, K., Marinov, P.: Generalized Nets and Ant Colony Optimization, Generalized Nets in Artificial Intelligence, vol. 5. Prof. Marin Drinov Academic Publishing House, Sofia (2011)

Chapter 2
Ant Colony Optimization

2.1 ACO Structure

Ants are found on all continents except Antarctica. Ant families are a perennial, highly organized community. Ants live in organized groups called colonies and have complex relationships with each other. Ants have an extremely strong sense of smell and are in contact with each other mainly by pheromones. The secrets are different—to mark the way to food, sexual secretions, danger signals, pheromones for type and service recognition, which they perform. From the point of view of computer science and of interest is the behavior of ants in the process of localization and assimilation of food sources for the survival needs of the colony as a whole. Beyond the behavior of the individual, individuals in the colony are found global rules for the behavior of anthill in general, which can be applied to specific areas in mathematics and informatics in the form of optimization algorithms. Ants use many pheromones in their communication. When any ant comes across a food source, it begins to secrete a pheromone, which in her way back to the anthill leaves a kind of smelly path to the food. All ants found food, leave on return this pheromone trail, forming something as an aromatic network pointing to food sources. When there is no pheromone trace, the ant takes in random direction, but in the presence of such, the ant most will probably take the road with the most intense pheromone trail. This results in an autocatalytic process in which ants detect the shortest paths to food sources. Over time, the pheromone evaporates and if not renewed, its quantity decreases. This prevents the possibility of the ants following some old trail pointing to an already exhausted one food source. In an experiment called the Double Bridge [1], the authors do research on how to place a pheromone and the relationship with the behavior of the ants. First Marco Dorigo applys ideas from ants behavior to solve combinatorial optimization problems [2–4].

Ant algorithms are part of metaheuristic methods for optimization. Metaheuristics is a high-level procedure designed to find, build or choose a low-level procedure that can ensure finding enough a good solution to an optimization problem, especially

S. Fidanova, *Ant Colony Optimization and Applications*, Studies in Computational Intelligence 947, https://doi.org/10.1007/978-3-030-67380-2_2

when the information is incomplete or computer resources are limited. The first ACO algorithm aimed to find the optimal path in a graph based on the behavior of ants looking for a way between their colony and a source of food. From then until today, Dorigo's original idea has been expanded and modified by different researchers to apply to a wider class of tasks. From then until today, Dorigo's original idea has been expanded and modified by different researchers to apply to a wider class of tasks.

2.1.1 Graph of the Problem

Ants in the nature exchange information indirectly by placing a pheromone in their path. This mechanism for solving a task, that is too complex for an individual, is a good example of a self-organizing system. This system is based on positive feedback (increasing the pheromone attracts other ants that amplify it) and negative feedback (decreasing the pheromone due to evaporation). Even slight changes in the pheromone allow a route to be preferred.

This behavior of ants gives the idea of presenting the problem with a graph, and the potential solutions as paths in the graph. The optimization problem comes down to finding the shortest path in a graph with additional constraints. Depending on the task itself, the pheromone can be placed on the arcs of the graph or on the nodes of the graph. The graph representation is one of the important points of the algorithm. One needs to decide which elements of the problem to be represented by the nodes, what is the meaning of the arcs.

2.1.2 Solution Construction

The main step in ant algorithm is solution construction. The ant algorithm can generally be represented as follows. The method is iterative. At each iteration, each of the ants begins to build its solution, starting from a random node in the graph. Random start is a diversification of the search in the space of the solutions. The ant chooses the next node to be included in the solution using a function called transition probability. This function is a product of the amount of pheromone, corresponding to this transition, and heuristic information. Heuristic information is problem dependent. The amount of pheromone represents the experience of ants from previous iterations. Heuristic information is a function representing prior knowledge of the task. The construction of heuristic information is essential for finding good solutions. This is usually an appropriate combination of the parameters of the objective function and the constraints of the task. The ant chooses to add to the solution the node highest transition probability. If there are two or more nodes with equal probability, then one of them is chosen at random.

The ant moves from node i to node j of the graph with probability:

$$p_{ij} = \frac{\tau_{ij}^{\alpha}\eta_{ij}^{\beta}}{\sum_{all\ possible\ k} \tau_{ik}^{\alpha}\eta_{ik}^{\beta}} \tag{2.1}$$

where

- τ_{ij} is the amount of pheromone corresponding to the transition from node i to node j;
- α is a parameter to control the influence of τ_{ij};
- η_{ij} is the heuristic information;
- β is a parameter to control the influence of η_{ij}.

2.1.3 *Pheromone Updating*

Ants in the nature place a pheromone in the path they follow so that they can return to the nest. Shorter pathways accumulate more pheromones and so over time they become more desirable and are followed by more ants.

In ACO algorithm our artificial ants mimic real ants. They deposit a pheromone on the elements of the graph (arcs or nodes), which are related with their solutions. At the beginning of the algorithm the pheromone is initialized to a small positive constant value $0 < \tau_0 < 1$. We can consider the pheromone τ_{ij} as past experience gained by the colony, as its global memory with respect to using arc (i, j) in past solutions.

When all ants have constructed their solutions, pheromone trails are updated with respect of the current experience. First, pheromone trails are "evaporated", each value τ_{ij} is multiplied by a factor $1 - \rho$, where $\rho \in [0, 1]$ is the evaporation rate. Thus ants partially forget older experience and the ants focus on new information. Then the ants add new pheromone. The quantity of the new added pheromone depends of the quality of the achieved solution. The elements of better solutions receive more pheromone than others and are more likely to be selected in the next iteration.

The pheromone updating rule is as follows:

$$\tau_{ij} = (1 - \rho)\tau_{ij} + \Delta\tau_{ij}, \tag{2.2}$$

where τ_{ij} is the amount of pheromone corresponding to the transition from node i to node j. ρ is the evaporation rate of the pheromone.

2.1.4 *Algorithm's Parameters*

ACO algorithm has several parameters, on which depends algorithm performance. They are:

- number N of used ants;
- evaporation rate ρ;
- initial pheromone τ_0;
- number I of iterations;
- α and β.

ACO is population based method, which uses colony of ants. In every iteration an ant begins solution construction from random node of the graph of the problem. The random start is a diversification of the search. Thanks to it, the algorithm uses a small population, 20 ants are enough. For comparison, Genetic Algorithm (GA) needs a population with minimum 100 individuals [5].

Evaporation parameter simulates evaporation process in the nature. It shows how important is the old information according the new one. Evaporation parameter is problem dependent. Its value is in the interval [0, 1].

The algorithm is not very sensitive to the initial pheromone value. The most important is to be strongly positive. If the initial pheromone is 0, than the transition probability will be 0 and the algorithm will stop before starting the first iteration.

ACO is an iterative process. One of the stopping criteria is the number of iterations. Other possible stopping criteria can be fixed number of iterations without improvement of the solutions.

The parameters α and β show importance of the pheromone quantity and heuristic information respectively in the transition probability function. They are problem dependent.

2.2 Variants of ACO Algorithm

There are exist various variants of ant algorithm. The main difference between them is the way the pheromone is updated. The aim is to avoid local minimums and to improve the algorithm convergence.

2.2.1 *Ant System*

The first variant of the ACO method is Ant System [6]. It follows exactly the behavior of ants in the wild. In this variant of the method all ants deposit a pheromone, proportional to the quality of their solution. The observations are that the elements of the graph accumulate too many pheromone and the algorithm stagnate in very early stage. Experimentally is found that it is inferior to other known metaheuristic methods. Therefore, various variants of the method have been created in order to improve the quality of solutions found.

2.2.2 *Elitist Strategy for Ant System*

This is the first modification of the algorithm, called the strategy of elite individuals, proposed by Dorigo [4]. In it, a fixed number of ants (one or more), achieving the best solution, are declared elite and only they get the right to put a pheromone on their route. The goal is to provide an opportunity to increase the pheromone values of those elements of the graph that have been found to be part of the best so far solutions. This is a technique for concentrating the search around already found good solutions, assuming that there are other good solutions close to them. If all ants are declared elite, then this option coincides with Ant System.

2.2.3 *Ant Colony System*

In Ant Colony System, the main is that there are two types of pheromone updates, global and local. In the global pheromone update, the pheromone of all elements of the graph that are part of solutions, constructed during the current iteration, is decreased (evaporated) and then a new pheromone, equal to the initial one, is added. In this way it is ensured that the amount of pheromone in the elements of the graph will be not less than the initial pheromone. In the local renewal of the pheromone, only the elements of the best solution receive a pheromone proportional to the value of the objective function. With the global updating of the pheromone, a diversification of the search is achieved. The possibility of the pheromone value approaching zero is prevented and leaving some areas of the set with solutions unexplored. The local updating of the pheromone enhances its value around the best solution found and concentrates the search near it.

2.2.4 *Max-Min Ant System*

Also known as MAX-MIN ACO or MMAS [7]. The algorithm is changed by fixing the upper and lower bounds of the pheromone $[\tau_{min}, \tau_{max}]$. The pheromone is deposited only on the elements of the graph, corresponding to the globally best solution or the best solution for the current iteration. The rest of the pheromone just evaporates. If the amount of pheromone in an element becomes less than the fixed lower bound, it is set to be equal to the lower bound. Thus, the amount of pheromone cannot be close to zero and the possibility of any area of the set with solutions remaining unexplored is prevented. If the amount of pheromone in an element becomes greater than the fixed upper bound, it is set to be equal to the upper bound. After the first iteration, all elements of the graph receive a pheromone equal to τ_{max}. These changes aim to have no elements in the graph that concentrate too much pheromone and this to leads to the repetition of the same solutions, stagnation of the search process. On the other hand, the pheromone of some elements is prevented from being close to zero and these elements to become undesirable.

2.2.5 ACO with Additional Reinforcement

In this variant of the method, the pheromone is received by the elements of the graph, belonging to the best solution [8]. The elements of the graph, which are not included in the decision of any ant, also receive a pheromone. The pheromone of the elements, belonging to the other solutions remains unchanged. Thus, on the one hand, the search is diversifying, with ants being forced to look for solutions in unexplored areas of the search space, avoiding those already found bad decisions. On the other hand, there is a concentration of search around the best found so far solution.

References

1. Deneubourg, J.L., Aron, S., Goss, S., Pasteels, J.M.: The self-organising exploratory pattern of the Argentine ant. J. Insect Behav. **3**, 159–168 (1990)
2. Dorigo, M., Di Caro, G.: The ant colony optimization meta-heuristic. In: Corne, D., Dorigo, M., Glover, F. (eds.) New Idea in Optimization, pp. 11–32. McGrow-Hill, London (1999)
3. Dorigo, M., Gambardella, L.M.: Ant colony system: a cooperative learning approach to the traveling salesman problem. IEEE Trans. Evol. Comput. **1**, 53–66 (1997)
4. Dorigo, M., Stutzle, S.: Ant Colony Optimization. MIT Press, Cambridge (2004)
5. Fidanova, S., Paprzycki, M., Roeva, O.: Hybrid GA-ACO algorithm for a model parameter identification problem. In: Proceeding of FedCSIS 2014 Conference, pp. 413–420. IEEE Xplorer (2014)
6. Bonabeau, E., Dorigo, M., Theraulaz, G.: Swarm Intelligence: From Natural to Artificial Systems. Oxford University Press, New York (1999)
7. Shmygelska, A., Hoos, H.H.: An ant colony optimization algorithm for the 2D and 3D hydrophobic polar protein folding problem. BMC Bioinf. **6**(30). https://doi.org/10.1186/1471-2105-6-30 (2005)
8. Fidanova, S.: ACO algorithm with additional reinforcement. Inter. Conf. Ant Colonies Artif. Ants, Lect. Notes Comput. Sci. **2463**, 292–293. Springer (2002)

Chapter 3
Multiple Knapsack Problem

The text of this chapter is based on [1–3]. The Multiple Knapsack problem (MKP) is a hard combinatorial optimization problem with large application, which embraces many practical problems from different domains, including transport, cargo loading, cutting stock, bin-packing, financial management, etc. It is an important part of more complex optimization problems like vehicle routing problem.

3.1 Problem Formulation

The Knapsack Problem has numerous applications in theory as well as in practice. It also arises as a subproblem in several algorithms for more complex Combinatorial Optimization Problems (COPs) and these algorithms will benefit from any improvement in the field of MKP. We can mention the following major applications: problems in cargo loading, cutting stock, bin-packing, budget control and financial management may be formulated as MKP. In [4] is proposed to use the MKP in fault tolerance problem and in [5] is designed a public cryptography scheme whose security realize on the difficulty of solving the MKP. Martello and Toth [6] mention that two-processor scheduling problems may be solved as a MKP. Other applications are industrial, naval, aerospace management, computational complexity theory, etc.

The shortest path problem in a transportation network deals with determining the subset of the connected roads that collectively comprise the shortest driving distance or the smallest driving time or the cheapest fair between two cities. The problem is what subset of lines gives the faster response time for communication between them. Complexity theory is part of the theory of computation dealing with the resources required during the computation to solve a given problem.

The more theoretical application either appear where a general problem is transformed to a MKP or where a MKP appears as a subproblem. We should mentioned that MKP appears as a subproblem when solving the generalized assignment

S. Fidanova, *Ant Colony Optimization and Applications*, Studies in Computational Intelligence 947, https://doi.org/10.1007/978-3-030-67380-2_3

problem, which again is heaving used when solving vehicle routing problems. In addition, MKP can be seen as a general model for any kind of binary problems with positive coefficients [7].

The MKP can be thought as a resource allocation problem, where we have m resources (the knapsacks) and n objects and object j has a profit p_j. Each resource has its own budget c_i (knapsack capacity) and consumption r_{ij} of resource i by object j. We are interested in maximizing the sum of the profits, while working with a limited budget.

The MKP can be formulated as follows:

$$\begin{array}{rl} \max & \sum_{j=1}^{n} p_j x_j \\ \text{subject to} & \sum_{j=1}^{n} r_{ij} x_j \leq c_i \;\; i = 1, \ldots, m \\ & x_j \in \{0, 1\} \;\; j = 1, \ldots, n. \end{array} \tag{3.1}$$

There are m constraints in this problem, so MKP is also called m-dimensional knapsack problem. Let $I = \{1, \ldots, m\}$ and $J = \{1, \ldots, n\}$, with $c_i \geq 0$ for all $i \in I$. A well-stated MKP assumes that $p_j > 0$ and $r_{ij} \leq c_i \leq \sum_{j=1}^{n} r_{ij}$ for all $i \in I$ and $j \in J$. Note that the $[r_{ij}]_{m \times n}$ matrix and $[c_i]_m$ vector are both non-negative.

The partial solution is represented by $S = \{i_1, i_2, \ldots, i_j\}$, where $x_{i_s} = 1 \quad (s = 1, \ldots, j)$, and the most recent element incorporated to S, i_j needs not be involved in the process of selecting the next element. One of the basic elements of the ACO metaheuristic is the mapping of the problem onto a graph, thus the solutions of the problem are represented by paths through the graph. We define the graph of the problem as follows: the nodes correspond to the objects, the arcs fully connect nodes. Fully connected graph mens that after the object i we can chose the object j, for every i and j if the object j is not chosen yet.

3.2 Heuristic Model

When we calculate the selection probability we can use various types of heuristic information. It can be static and dynamic, and we can use different problems' parameters for heuristic. Static heuristic stays unchanged during the run of the algorithm, while the dynamic heuristic corresponds to the current state of the problem. Obviously, we will include the profit in the heuristics because it is the most important information for objective function. We expect that we will achieve better results, if we include more information for the problem in heuristic. Thus we will try to use the constraints in different manners and will compare the results.

3.2.1 Static Heuristics

We propose two types of static heuristics using constraints. We will call them "heuristic A" and "heuristic B" respectively.

- **Heuristic A**
  Let $s_j = \sum_{i=1}^{m} r_{ij}$. For heuristic information we use:

$$\eta_{ij} = p_j^{d_1} / s_j^{d_2}, \tag{3.2}$$

  $0 < d_1$ and $0 < d_2$ are parameters. We include the expenses in heuristic. Hence the objects with greater profit and less average expenses will be more desirable. Thus we try to do some balance between expenses and the profit for a given object.
- **Heuristic B**
  Let $s_j = \sum_{i=1}^{m} r_{ij}/c_i$. For heuristic information we use:

$$\eta_{ij} = p_j^{d_1} / s_j^{d_2}, \tag{3.3}$$

  $0 < d_1$ and $0 < d_2$ are parameters. Thus the heuristic depends by the profit, the expenses and the budgets. The objects with greater profit which use less parts of the budgets will be more desirable.

3.2.2 Dynamic Heuristic

The third and the forth types of heuristics are dynamic and they correspond to the current state of the algorithm. We will call them "heuristic C" and "heuristic D" respectively.

- **Heuristic C** [8]
  Let $b_i = c_i - \sum_{j=1}^{n} r_{ij} x_j$ is the remainder of the budget before choosing the next object, and $s_j = \sum_{i=1}^{m} r_{ij}/b_i$ if $b_i \neq 0$ and $s_j = \sum_{i=1}^{m} r_{ij}$ if $b_i = 0$. For heuristic information we use:

$$\eta_{ij} = p_j^{d_1} / s_j^{d_2}, \tag{3.4}$$

  where $d_1 = d_2$. The aim is the heuristic information to have maximal correspondence to the current state of the algorithm and thus to achieve good result. In [8] the authors do not verify if $b_i \neq 0$, but because it can happen and there is division by b_i, we add this verification in the algorithm.

- **Heuristic D**
  For heuristic information we use:

$$\eta_{ij} = p_j^{d_1} / s_j^{d_2}, \tag{3.5}$$

  $0 < d_1$ and $0 < d_2$ are parameters and they can be different. The mining of s_j is like in heuristic C. Thus the Heuristic C is a particular case of Heuristic D.

The parameters d_1 and d_2 show the importance of the profit and the constraints in heuristic information. If $d_1 > d_2$, than the profit is more important and in opposite case the constraints are more important. If we use greater values for d_1 and d_2, than the heuristic information is more important than the pheromone in the transition probability.

3.2.3 Comparison of Different Heuristic Models

We were running experiments of our ACO algorithm for all four heuristics on 5 different instances from "OR-Library" available within WWW access at http://mscmga.ms.ic.ac.uk/jeb/orlib, with 100 objects and 10 constraints. For all 5 instances we were running experiments for a range of evaporation rates and the parameters d_1 and d_2 in order to find the best rate for every instance. The results of this experiments are shown in Fig. 3.1. We fixed the initial pheromone value to be $\tau_0 = 0.5$. The developed technique has been coded in C++ language and implemented on a Pentium III (450 MHz). Because of the random start of the ants in every iteration we can use less ants than the number of objects. After the tests we found that 10 ants are enough to achieve good result for a reasonable time. After choosing for every problem instance the best rate for parameters we could compare the different heuristic representations. In Fig. 3.1 we show average results over all 5 problem instances and every instance is run 30 times with the same parameter settings. To can observe the influence of the heuristics better, we use ACO algorithm without local search or other methods to improve the results and we run the different test instances on same random sequences. First our observation is that the heuristic B shows advantage over the other three heuristics representations. It means that it is important the chosen objects to have small expenses, but it is more important the expenses to be a small part of the relevant budget. We expected to achieve better results by dynamic heuristics because they correspond to the current state of the problem. In spite of our expectations we achieve weaker results by dynamic heuristics. Comparing heuristics C and D we observe the importance of the parameters d_1 and d_2. In the case $d_1 \neq d_2$ the achieved results are better. Nevertheless we do not use any method to improve the results we achieve results close to the best found in the literature.

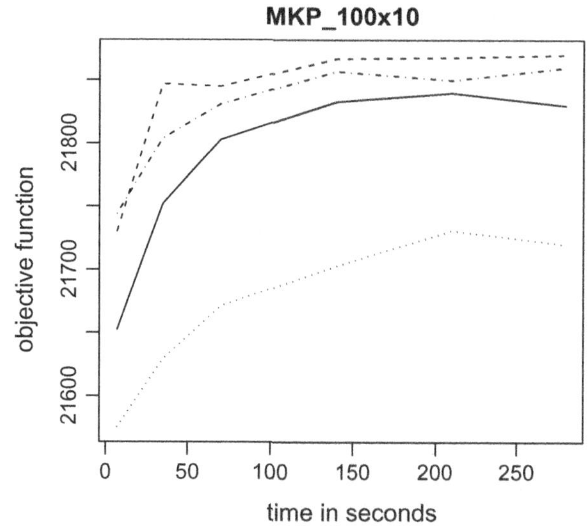

Fig. 3.1 The graphics shows the average solution quality (value of the total cost of the objects in the knapsack) over 30 runs. Dash dot line represents heuristic A, dash line—heuristic B, dots line—heuristic C and thick line—heuristic D

3.3 Pheromone Model

There are two possibilities for placing a pheromone. It can be placed on the arcs of the graph of the problem or on the nodes like in [8]. In the case of arcs τ_{ij}, in the formulas, corresponds to the pheromone quantity on the arc (i, j). In the case of nodes τ_{ij} corresponds to the pheromone quantity on the node j.

To explain the difference between the two proposed pheromone models we consider a small problem instance with $n = 5$ objects and one constraint.

$$\begin{aligned} &max(5x_1 + 5x_2 + 4x_3 + 6x_4 + 6x_5) \\ &5x_1 + 2x_2 + 3x_3 + 4x_4 + 3x_5 \leq 10 \end{aligned} \tag{3.6}$$

The optimal solution of this problem is (01011) and the optimal value of the object function is $L_{opt} = 17$. One ant will be used and it will start from node 1 of the graph of the problem in the first iteration and from nodes 3, 2 and 5 in the second, third and fourth iterations respectively, for both pheromone models. In the case of pheromone on the nodes, the achieved solutions are (11001) $L = 16$, (01101) $L = 15$, (11001) $L = 16$, (11001) $L = 16$. We observe that in iterations 3 and 4 the ant starts from nodes belonging to the optimal solution but it can not find the optimal one. In the case of pheromone on the arcs, the achieved solutions are (11001) $L = 16$, (01101) $L = 15$, (01011) $L = 17$, (01011) $L = 17$. In this case we observe that when the ant starts from the node belonging to the optimal solution, it finds the optimal one.

When the pheromone is on the nodes some of them accumulates large amount of pheromone and become much more desirable then all others. In the other case this pheromone is dispersed on the arcs that enter the node.

We show the computational experience of the ACO using MKP from "OR-Library" available within WWW access at http://mscmga.ms.ic.ac.uk/jeb/orlib. To provide a fair comparison for the above implemented ACO algorithm, a predefined number of iterations, K = 500, is fixed for all the runs. The developed technique has been coded in C++ language and implemented on a Pentium 3 (450 MHz). Because of the random start of the ants we can use less ants then the number of the nodes (objects). After the tests we found that 10 ants are enough to achieve good results. Thus we decrease the running time of the program. The results on the Figs. 3.2 and 3.3 show the advantage of putting the pheromone on the arcs. The reported results are average value of the objective function over 30 runs of the program with different number of iterations. For all tested problems our algorithm achieves, in some of the runs, the best results found in the literature.

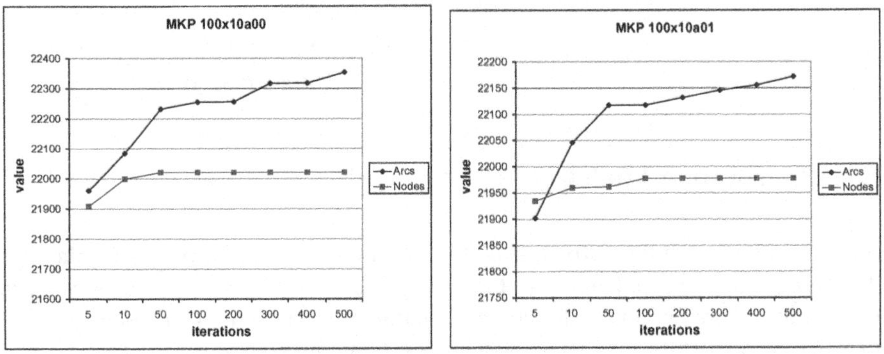

Fig. 3.2 The figures shows the average solution quality (value of the total cost of the objects in the knapsack) per iteration over 30 runs

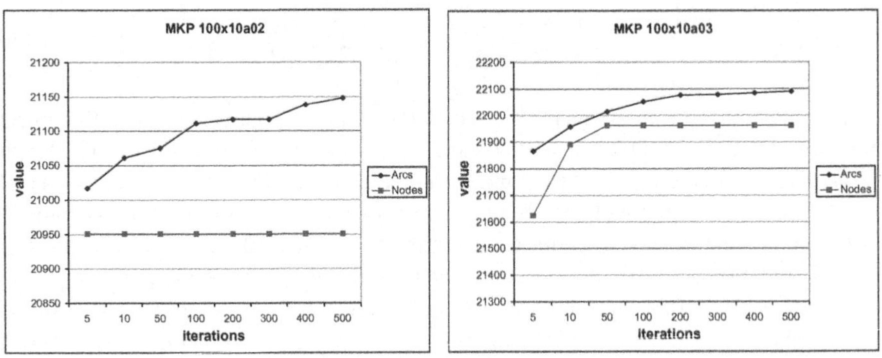

Fig. 3.3 The figures shows the average solution quality (value of the total cost of the objects in the knapsack) per iteration over 30 runs

We observe that we have very early stagnation of the algorithm in the case of pheromone on the nodes. We can explain this appearance with a large pheromone accumulation on some nodes and thus ants repeat same solution over and over again. In the case of pheromone on the arcs the difference between the amount of the pheromone on the arcs is not so big.

3.4 Probabilistic Model

In this section we describe four possibilities for transition probability model. For ant k, the probability p_{ij}^k of moving from a state i to a state j depends on the combination of two values:

- The attractiveness η_{ij} of the move as computed by some heuristic.
- The pheromone trail level of the move.

The pheromone τ_{ij} is associated with the arc between nodes i and j.

3.4.1 *Proportional Transition Probability*

The quantity of the pheromone on the arcs between two nodes is proportional to the experience of having the two nodes in the solution. Thus the node j is more desirable if the quantity of the pheromone on arc (i, j) is high. For ant k which moves from node i to node j the rule is:

$$p_{ij}^k(t) = \begin{cases} \frac{\tau_{ij}\eta_{ij}(S_k(t))}{\sum_{q \in allowed_k(t)} \tau_{iq}\eta_{iq}(S_k(t))} & \text{if } j \in allowed_k(t) \\ 0 & \text{otherwise} \end{cases}, \tag{3.7}$$

- $allowed_k$ is the set of remaining feasible states.

3.4.2 *Transition Probability with Sum*

This probability takes into account how desirable in the past has been the node j, independently how many ants have reached it by the node i or by some other node. Thus the node j is more desirable if the average quantity of the pheromone on the arcs which entry in the node j is high. In this case the transition probability becomes:

$$p_{ij}^k(t) = \begin{cases} \frac{(\sum_{i=1}^n \tau_{ij})\eta_{ij}(S_k(t))}{\sum_{q \in allowed_k(t)} (\sum_{i=1}^n \tau_{iq})\eta_{iq}(S_k(t))} & \text{if } j \in allowed_k(t) \\ 0 & \text{otherwise} \end{cases} . \tag{3.8}$$

3.4.3 Maximal Transition Probability

This probability is proportional to the maximal pheromone on the arcs which entry in the node j. Thus the node j will be more desirable, independently of the quantity of the pheromone on the arc (i, j), if there is some other arc with high quantity of the pheromone which entry in the node j. In this case the transition probability is changed as follows:

$$p_{ij}^k(t) = \begin{cases} \frac{(max_i \tau_{ij})\eta_{ij}(S_k(t))}{\sum_{q \in allowed_k(t)} (max_i \tau_{iq})\eta_{iq}(S_k(t))} & \text{if } j \in allowed_k(t) \\ 0 & \text{otherwise} \end{cases} . \tag{3.9}$$

3.4.4 Minimal Transition Probability

This probability is proportional to the minimal pheromone on the arcs which entry in the node j. Thus the node j will be more desirable if the quantity of the pheromone on all arcs which entry in the node j is high. In this case the transition probability is as follows:

$$p_{ij}^k(t) = \begin{cases} \frac{(min_i \tau_{ij})\eta_{ij}(S_k(t))}{\sum_{q \in allowed_k(t)} (min_i \tau_{iq})\eta_{iq}(S_k(t))} & \text{if } j \in allowed_k(t) \\ 0 & \text{otherwise} \end{cases} . \tag{3.10}$$

3.4.5 Experimental Results

In this subsection we describe the experimental analysis on the performance of MKP as a function of the transition probability. We show the computational experience of the ACO using 10 MKP instances from "OR-Library" available within WWW access at http://people.brunel.ac.uk/~mastjjb/jeb/orlib, with 100 objects and 10 constraints. To provide a fair comparison for the above implemented ACO algorithm, a predefined number of iterations, $k = 500$, is fixed for all the runs. The developed technique has been coded in C++ language and implemented on a Pentium 4 (2.8 GHz).

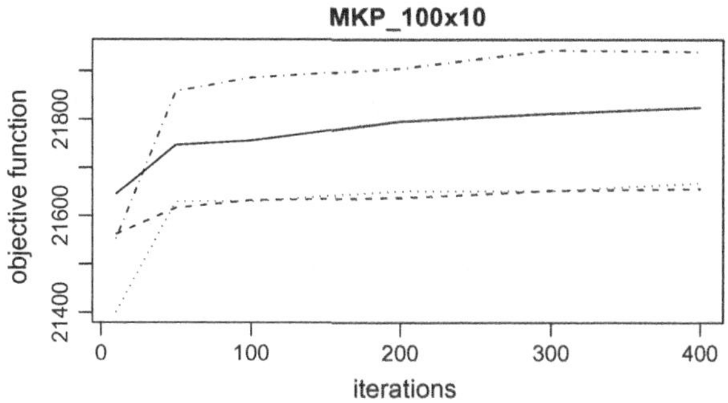

Fig. 3.4 The graphics shows the average solution quality (value of the total cost of the objects in the knapsack) over 30 runs. Dash dot line represents proportional probability, dash line—probability with sum, dots line—maximal probability and thick line—minimal probability

Because of the random start of the ants in every iteration, we can use less ants than the number of the nodes. After the tests we found that 10 ants are enough to achieve good results. Thus we decrease the running time of the program. We run the same instance using different transition probability models on the same random sequences for starting nodes and we find different results. Thus we are sure that the difference comes from the transition probability. For all 10 instances we were running experiments for a range of evaporation rates and the parameters d_1 and d_2 in order to find the best parameters for every instance. We fixed the initial pheromone value to be $\tau_0 = 0.5$. After choosing for every problem instance the best rate for the parameters we could compare the different transition probabilities. In Fig. 3.4 we show the average result over all 10 problem instances and every instance is run 30 times with the same parameter settings.

First our observation is that the proportional and the minimal transition probabilities show advantage over the sum and maximal transition probabilities. In a small number of iterations (less than 50), probability with sum and minimal probability achieve better results, but after that the proportional probability outperforms them. The MKP is not ordered problem. It means that the quality of the solution is not related to the order we choose its elements. Using maximal transition probability it is enough only one arc to have high quantity of the pheromone and the node will be more desirable. This kind of probability is more suitable to ordered problems: the node j is more desirable by node i than by node q. Using transition probability with sum the node is more desirable if the average quantity of the pheromone is high, but for some of the arcs this quantity can be very high and for other arcs it can be very low. Thus we can explain the worst results with this two models of the transition probability. If the minimal probability is high, the quantity of pheromone for all arcs which entry the node is high. If the proportional probability is high it means that after node i is good to chose node j. The last two models of transition probability are better related to the unordered problems and thus we can achieve better results using them.

3.5 Conclusion

The design of a meta-heuristic, including ant colony optimization, is a difficult task and highly dependent on the structure of the optimized problem.

We have presented different types of heuristic informations. When we use profit of the objects and constraints we obtain better results than using only the profit of the objects. The results indicate the potential of the ACO approach for solving multiple knapsack problem. We can use similar technique for another constraint problems too. Our empirical results show that our ACO algorithm is currently among the best performing algorithms for this problem [8, 9].

Two models of the pheromone placement have been proposed, on the arcs and on the nodes. The comparison of the performance of the ACO coupled with these pheromone models applied to different MKP problems are reported. The obtained results show that when the pheromone is on he arcs, the the algorithm achieves better results, because it better corresponds to the structure of the problem.

Four models of the transition probability have been proposed. The comparison of the performance of the ACO coupled with these probability models applied on different MKP problems is reported. The goal is to find probability model which is more relevant to the structure of the problem.

References

1. Fidanova, S.: ACO Algorithm for MKP using various heuristic information. Numer. Methods Appl., Lect. Notes Comput. Sci. **2542**, 434–440. Springer (2003)
2. Fidanova, S.: Ant colony optimization for multiple knapsack problem and model bias. Numer. Anal. Appl., Lect. Notes Comput. Sci. **3401**, 282–289. Springer (2005)
3. Fidanova, S.: Heuristics for multiple knapsack problem. IADIS Applied Computing 2005 Conference, Algavre, pp. 255–260 (2005)
4. Sinha, A., Zoltner, A.A.: The multiple-choice knapsack problem. J. Oper. Res. **27**, 503–515 (1979)
5. Diffe, W., Hellman, M. E.: New direction in cryptography. IEEE Trans, Inf. Theory **36**, 644–654 (1976)
6. Martello, S., Toth, P.: A mixtures of dynamic programming and branch-and-bound for the subset-sum problem. Manag. Sci. **30**, 765–771 (1984)
7. Atanassov, K., Atanassova, V., Gluhchev, G.: InterCriteria analysis: ideas and problems. Notes on Intuitionistic Fuzzy Sets **21**(1), 81–88 (2015)
8. Leguizamon, G., Michalewicz, Z.: A new version of ant system for subset problems. In proceedings of the Congress on Evolutionary Computing, pp. 1459–1464 (1999)
9. Chu, P.C., Beasley, J.E.: A genetic algorithm for the multiple knapsack problem. J. Heuristics **4**, 63–86 (1998)

Chapter 4
Grid Sheduling Problem

The text of this chapter is based on [1]. Grid computing is a form of distributed computing that involves coordinating and sharing computing, application, data storage or network resources across dynamic and geographically dispersed organizations. The goal of grid task scheduling is to achieve high system throughput and to match the application needed with the available computing resources. This is matching of resources in a non-deterministically shared heterogeneous environment. The complexity of scheduling problem increases with the size of the grid and becomes highly difficult to solve effectively. To obtain good methods to solve this problem a new area of research is implemented. This area is based on developed heuristic techniques that provide an optimal or near optimal solution for large grids. In this chapter we introduce a tasks scheduling algorithm for grid computing. The algorithm is based on Ant Colony Optimization. The chapter shows how to search for the best tasks scheduling for grid computing in dynamik way.

4.1 Introduction

Computational Grids are a new trend in distributed computing systems. They allow the sharing of geographically distributed resources in an efficient way, extending the boundaries of what we perceive as distributed computing. Various sciences can benefit from the use of grids to solve CPU-intensive problems, creating potential benefits to the entire society. With further development of grid technology, it is very likely that corporations, universities and public institutions will exploit grids to enhance their computing infrastructure. In recent years there has been a large increase in grid technologies research, which has produced some reference grid implementations.

Task scheduling is an integrated part of parallel and distributed computing. Intensive research has been done in this area and many results have been widely accepted. With the emergence of the computational grid, new scheduling algorithms are in

S. Fidanova, *Ant Colony Optimization and Applications*, Studies in Computational Intelligence 947, https://doi.org/10.1007/978-3-030-67380-2_4

demand for addressing new concerns arising in the grid environment. In this environment the scheduling problem is to schedule a stream of applications from different users to a set of computing resources to maximize system utilization. This scheduling involves matching of applications needs with resource availability. There are three main phases of scheduling on a grid [2]. Phase one is resource discovery, which generates a list of potential resources. Phase two involves gathering information about those resources and choosing the best set to match the application requirements. In the phase three the job is executed, which includes file staging and cleanup. In the second phase the choice of the best pairs of jobs and resources is NP-complete problem [3]. A related scheduling algorithm for the traditional scheduling problem is Dynamic Level Scheduling (DLS) algorithm [4]. DLS aims at selecting the best subtask-machine pair for the next scheduling. To select the best subtask-machine pair, it provides a model to calculate the dynamic level of the task-machine pair. The overall goal is to minimize the computational time of the application. In the grid environment the scheduling algorithm no longer focuses on the subtasks of an application within a computational host or a virtual organization (clusters, network of workstations, etc.). The goal is to schedule all the incoming applications to the available computational power. In [5, 6] some simple heuristics for dynamic matching and scheduling of a class of independent tasks onto a heterogeneous computing system have been presented. There are two different goals for task scheduling: high performance computing and high throughput computing. The former aims is minimizing the execution time of each application and later aims is scheduling a set of independent tasks to increase the processing capacity of the systems over a long period of time. Our approach is to develop a high throughput computing scheduling algorithm based on ACO. ACO algorithm can be interpreted as parallel replicated Monte Carlo (MC) systems [7]. MC systems [8] are general stochastic simulation systems, that is, techniques performing repeated sampling experiments on the model of the system under consideration by making use of a stochastic component in the state sampling and/or transition rules. Experimental results are used to update some statistical knowledge about the problem, as well as the estimate of the variables the researcher is interested in. In turn, this knowledge can be also iteratively used to reduce the variance in the estimation of the described variables, directing the simulation process toward the most interesting state space regions. Analogously, in ACO algorithms the ants sample the problem's solution space by repeatedly applying a stochastic decision policy until a feasible solution of the considered problem is built. The sampling is realized concurrently by a collection of differently instantiated replicas of the same ant type. Each ant "experiment" allows to adaptively modify the local statistical knowledge on the problem structure. The recursive retransmission of such knowledge determines a reduction in the variance of the whole search process the so far most interesting explored transitions probabilistically bias future search, preventing ants to waste resources in not promising regions of the search.

4.2 Grid Scheduling Model

Our scheduling algorithm is designed for distributed systems shared asynchronously by both remote and local users.

4.2.1 Grid Model

The grid considered in this study is composed of a number of hosts send, each host is composed of several computational resources, which may be homogeneous or heterogeneous. The grid scheduler does not own the local hosts, therefore does not have control over them. The grid scheduler must make best effort decisions and then submit the jobs to the hosts selected, generally as a user. Furthermore, the grid scheduler does not have control over the set of jobs submitted to the grid, or local jobs submitted to the computing hosts directly. This lack of ownership and control is the source of many of the problems yet to be solved in this area. The GRID scheduling is a particular case of tasks scheduling on machines problem. In the GRID scheduling every machine can execute any task, but for different time.

4.2.2 Grid Scheduling Algorithm

While there are scheduling request from applications, the scheduler allocates the application to the host by selecting the best match from the pool of applications and pool of the available hosts. The selecting strategy can be based on the prediction of the computing power of the host [9]. We will review some terms and definitions [6, 10].

The expected execution time ET_{ij} of task t_i on machine m_j is defined as the amount of time taken by m_j to execute t_i given that m_j has no load when t_i is assigned. The expected completion time CT_{ij} of the task t_i on machine m_j is defined as the wall-clock time at which m_j completes t_i (after having finished any previously assigned tasks). Let M be the total number of the machines. Let S be the set containing the tasks. Let the beginning time of t_i be b_i. From the above definitions, $CT_{ij} = b_i + ET_{ij}$. The makespan for the complete schedule is then defined as $\max_{t_i \in S}(CT_{ij})$. Makespan is a measure of the throughput of the heterogeneous computing system. The objective of the grid scheduling algorithm is to minimize the makespan. It is well known that the problem of deciding on an optimal assignment of jobs to resources is NP-complete. We develop heuristic algorithm based on ACO to solve this problem.

Existing mapping heuristics can be divided into two categories: on-line mode and batch mode. In the on-line mode, a task is mapped onto a machine as soon as it arrives at the mapper. In the batch mode, tasks are not mapped onto the machines as they arrive, instead they are collected in a set that is examined for mapping at pre-scheduled times called mapping events. This independent set of tasks that is considered for mapping at mapping events is called meta-task. In the on-line mode, each task is considered only once for matching and scheduling. The minimum completion time heuristic assigns each task to the machine so that the task will have the earliest computation time [11]. The minimum execution time heuristic assigns each task to the machine that performs that tasks' computation in the least amount of execution time. In batch mode, the scheduler consider a meta-task for matching and scheduling at each mapping event. This enable the mapping heuristics to possibly make better decision, because the heuristics have the resource requirement information for the meta-task and known the actual execution time of a larger number of tasks. Our heuristic algorithm is for batch mode.

Let the number of the tasks in the set of tasks is greater than the number of machines in the grid. The result will be triples $(task, machine, startingtime)$. The function $free(j)$—shows when the machine m_j will be free. If the task t_i is executed on the machine m_j then the starting time of t_i becomes $b_i = free(j) + 1$ and the new value of the function $free(j)$ becomes $free(j) = b_i + ET_{ij} = CT_{ij}$.

An important part of implementation of ACO algorithm is the graph of the problem. We need to decide which elements of the problem to correspond to the nodes and which ones to the arcs. Let $M = \{m_1, m_2, \ldots, m_m\}$ is the set of the machines and $t = \{t_1, t_2, \ldots, t_s\}$ is the set of the tasks and $s > m$. Let $\{T_{ij}\}_{s\times m}$ is the set of the nodes of the graph and to machine $m_j \in M$ corresponds a set of nodes $\{T_{kj}\}_{k=1}^{s}$. The graph is fully connected. The problem is to choose s nodes of the graph thus to minimize the function $F = max(free(j))$, where $[b_i, CT_{ij}] \cap [b_k, CT_{kj}] = \oslash$ for all $i,\ j,\ k$. We will use several ants and every ant starts from random node to create their solution. There is a tabu list corresponding to every ant. When a node T_{ij} is chosen by the ant, the nodes $\{T_{ik}\}_{k=1}^{m}$ is included in tabu list. Thus we prevent the possibility the task t_i to be executed more than ones. An ant add new nodes in the solution till all nodes are in the tabu list. Like heuristic information we use:

$$\eta_{ij} = \frac{1}{free(j)}.$$

Thus if a machine is free earlier, the corresponding node will be more desirable. At the end of every iteration we calculate the objective function $F_k = max(free(j))$ over the solution constructed by ant k and the added pheromone by the ant k is:

$$\Delta\tau_{ij} = \frac{(1-\rho)}{F_k}.$$

Hence in the next iterations the elements of the solution with less value of the objective function will be more desirable. Our ACO implementation is different to ACO implementation on traditional tasks machines scheduling problem. The new of our implementation is using of multiple node corresponding to one machine. It is possible because in GRID scheduling problem every machine can execute any task.

Two kind of sets of tasks are needed: set of scheduled tasks and set of arrived and unscheduled tasks. When the set of scheduled tasks becomes empty the sched-

Table 4.1 Makespan for the execution on first free machine and ACO algorithm

Online-mode (%)	ACO	Improvement (%)
80	67	16
174	128	26.4
95	80	15.8

uled algorithm is started over the tasks from the set of unscheduled tasks. Thus is guaranteed that the machines will be fully loaded.

4.3 Experimental Testing

We have developed 3 simulated grid examples to evaluate the newly proposed ACO algorithm for grid scheduling. In our experimental testing we use 5 heterogeneous machines and 20 tasks. The initial parameters are set as follows: $\tau_0 = 0.01$ and $\rho = 0.5$ and we use 1 ant. We compare achieved by ACO algorithm result with often used online-mode (Table 4.1).

The results are in minutes. We observe the outperform of ACO algorithm and the improvement of the result with. In online-mode the arriving order is vary important. In ACO algorithm the most important is the execution time of the separate task.

4.4 Conclusion

To confront new challenges in tasks scheduling in a grid environment, we present in this study heuristic scheduling algorithm. The proposed scheduling algorithm is designed to achieve high throughput computing in a grid environment. This is a NP-problem and to be solved needs an exponential time. Therefore the heuristic algorithm which finds a good solution in a reasonable time is developed. In this paper heuristic algorithm based on ACO method is discussed and it basic strategies for a grid scheduling are formulated. This algorithm guarantee good load balancing of the machines. In ACO technique it is very important how the graph of the problem is created. Another research direction is to create different heuristic based algorithms for problems arising in grid computing.

References

1. Fidanova, S.: Durchova, M.: Ant algorithm for grid sheduling problem. Large Scale Comput., Lect. Notes Comput. Sci. **3743**, 405–412. Springer (2006)

2. Schopf, J.M: A general architecture for scheduling on the grid, special issue of JPDC on grid computing (2002)
3. Fernandez-Baca, D.: Allocating modules to processors in a distributed system. IEEE Trans. Softw. Eng. **15**(11), 1427–1436 (1989)
4. Sih, G.C., Lee, E.A.: A compile-time scheduling heuristic for inter connection-constrained heterogeneous processor architectures. IEEE Trans. Parallel Distrib. Syst. **4**, 175–187 (1993)
5. Braun, T.D., Siegel, H.J., Beck, N., Bolony, L., Maheswaram, M., Reuther, A.I., Robertson, J.P., Theys, M.D., Jao, B.: A taxonomy for describing matching and scheduling heuristics for mixed-machine heterogeneous computing systems. IEEE Workshop on Advances in Parallel and Distributed Systems, pp. 330–335 (1998)
6. Maheswaran, M., Ali, S., Siegel, H.J., Hensgen, D., Freund, R.: Dynamic mapping of a class of independent tasks onto heterogeneous computing systems. In: 8th IEEE Heterogeneous Computing Workshop (HCW'99), pp. 30–44. San Juan, Puerto Rico (1999)
7. Strelsov, S., Vakili, P.: Variance reduction algorithms for parallel replicated simulation of uniformized Markov chains. J. discrete Event Dyn. Syst. Theory Appl. **6**, 159–180 (1996)
8. Rubinstein, R.Y.: Simulation and the Monte Carlo Method. John Wiley & Sons (1981)
9. Gong, L., Sun, X.H., Waston, E.: Performance modeling and prediction of non-dedicated network computing. IEEE Trans. Comput. **51**(9), 1041–1059 (2002)
10. Pinedo, M.: Scheduling: Theory, Algorithms and Systems. Prentice Hall, Englewood Clifts, NJ (1995)
11. Freund, R.F., Gherrity, M., Ambrosius, S., Camp-Bell, M., Halderman, M., Hensgen, D., Keith, E., Kidd, T., Kussow, M., Lima, J.D., Mirabile, F., Moore, L., Rust, B., Siegel, H.J.: Scheduling resources in multi-user heterogeneous computing environments with SmartNet. In: IEEE Heterogeneous Computing Workshop, pp. 184–199 (1998)

Chapter 5
ACO for GPS Surveying Problem

The text of this chapter is based on [1–8]. Modern space technology prove to be valuable tools for the solution of many real-life problems in fields such as weather forecasting, managing effects of natural and man-made disasters, as well as in surveying, mapping and geodesy. Therefore, the need to develop and optimize the use of a satellite system in an efficient manner, which can provide safer and more reliable navigation service for civil use is necessary. The Global Positioning System (GPS) networks need a regular surveys consisting of temporary placing GPS receivers sequentially at pre-chosen points. This paper aims to provide new and potentially powerful metaheuristic algorithms to develop and improve this space application. In designing GPS surveying network, a given set of earth points must be observed consecutively (schedule). The cost of the schedule is the sum of the time needed to go from one point to another. The problem is to search for the best order in which this observation is executed. Minimizing the cost of the schedule is the goal of this work.

Global Positioning System (GPS) plays an important role in many navigation systems produced today, and is beginning to play an increasing role in providing accurate time signals for many industries. Applications include land surveying, autonomous vehicle control including the smart highway system, marine navigation, air traffic control. The service is always available at any time and under any weather condition. In addition, satellite navigation systems have an impact in geoscience, in particular on surveying work in quick and effective determining positions and changes in positions networks. Measuring requires that the survey crew physically passes through all the intervening terrain to measure the distance between any two adjacent points. Surveying methods have undergone a revolutionary change over the last few years with the deployment of satellite navigation systems. The most widely known space systems are: the American NAVSTAR global positioning system, the Russian GLObal Navigation Satellite System (GLO NASS), and the European satellite navigation system (GALILEO).

S. Fidanova, *Ant Colony Optimization and Applications*, Studies in Computational Intelligence 947, https://doi.org/10.1007/978-3-030-67380-2_5

A survey design consists of a list of sessions to be observed. The cost of sessions is the cost of moving receivers between points. For small networks, optimal (cheapest) solutions are possible to be found, however for large networks near optimal solutions have to be accepted. There is a need of sophisticated algorithms for accurately and reliable processing the GPS signals. In this chapter, we investigate the use of GPS to establish surveying networks. GPS satellites continuously transmit radio signals to the Earth while orbiting it. A receiver, with unknown position on Earth, has to detect and converts the signals received from all of the satellites into useful measurements. These measurements would allow a user to compute a three-dimensional coordinate position: location of the receiver.

Solving this problem to optimality requires a very high computational time. Therefore, new methods are needed to provide near-optimal solutions for large networks within acceptable amount of computational effort. These techniques are usually based on structured metaheuristics [2, 9, 10].

5.1 Problem Formulation

The GPS network can be defined as set of stations $(a_1, a_2, \ldots, a_n)$, which are co-ordinated by placing receivers $(X1, X2, \ldots)$ on them to determine sessions $(a_1a_2, a_1a_3, a_2a_3, \ldots)$ between them. The problem is to search for the best order in which these sessions can be organized to give the best schedule. Thus, the schedule can be defined as a sequence of sessions to be observed consecutively. The solution is represented by linear graph with weighted edges. The nodes represent the stations and the edges represent the moving cost. The objective function of the problem is the cost of the solution which is the sum of the costs (time) to move from one point to another one, $C(V) = \sum C(a_i, a_j)$, where a_ia_j is a session in solution V. For example if the number of points (stations) is 4, a possible solution is $V = (a_1, a_3, a_2, a_4)$ and it can be represented by linear graph $a_1 \rightarrow a_3 \rightarrow a_2 \rightarrow a_4$. The moving costs are as follows: $C(a_1, a_3), C(a_3, a_2), C(a_2, a_4)$. Thus the cost of the solution is $C(V) = C(a_1, a_3) + C(a_3, a_2) + C(a_2, a_4)$.

In practice, the task is to determine how each GPS receiver should be moved between stations to be surveyed in an efficient manner taking into account some important factors such as time, cost etc. The problem is to search for the best order, with respect to the time, in which these sessions can be observed to give the cheapest schedule or to minimize $C(V)$. The initial data is a cost matrix, which represents the cost of moving a receiver from one point to another. The cost could be evaluated purely upon the time or purely upon the distance; for more details see Dare [9].

This chapter addresses the problem of determining the optimal session schedule, according to some cost criteria. If the cost of moving between two points is independent of the direction than a symmetric cost matrix will be produced, but the realistic case is non-symmetric and the problem becomes more difficult. In surveying, a non-symmetric cost matrix is more realistic as movements between the points usually require a combination of driving and walking and hence uphill journeys are usually slower than downhill journeys. If a helicopter is used to move the surveyors between the points, then a symmetric cost matrix may be more appropriate. A reconnaissance is usually carried out before the actual survey and it is at this time that data to enable costs to be calculated can be collected.

If S represents the number of the sessions, then the number of possible session schedules is given by $S!$, which is clearly a very large number for some networks. Therefore it is impossibles to find optimal solution for large networks in a reasonable time. In this case are applied methods from operational research (OR), which find quickly near optimal solutions.

The purpose of the present chapter is to show how ACO can be used to solve GPS sessions to be observed, given the cost to move receivers between points in the network. The session schedule is defined as a sequence of sessions to be observed consecutively. ACO is a technique to determine near-optimal solutions in a reasonable amount of computational resources. It is used to solve large problems that cannot be solved optimally and reasonably quickly.

5.2 ACO Algorithm for GPS Surveying Problem

In the proposed GPS-ACO technique an initialisation phase takes place during which ants are positioned on different nodes (sessions) with empty tabu lists and initial pheromone distributed equally on paths connecting these sessions. Ants update the level of pheromone while they are constructing their schedules by iteratively adding new sessions to the current partial schedule. At each time step, ants compute a set of feasible moves and select the best one according to some probabilistic rules based on the heuristic information and pheromone level. The higher value of the pheromone and the heuristic information, the more profitable is to select this move and resume the search. The selected node is putted in the tabu list related to the ant to prevent to be chosen again. Heuristic information represents the nearer sessions around the current session, while pheromone level memory of each path represents the usability of this path in the past to find good schedules. At the end of each iteration, the tabu list for each ant will be full and the obtained cheapest schedule is computed and memorized. For the following iteration, tabu lists will be emptied, ready for use and the pheromone level will be updated. This process is repeated till the number of iterations (stopping criteria) has been reached. In more details, the proposed GPS-ACO technique constructs the cheapest observation schedule for a GPS network.

It is important when implementing the proposed technique to provide suitable and carefully chosen structural and control components according to the size and type of the applied GPS network. The structural elements determine the procedure in which the GPS network problem is modeled in order to fit into the ACO framework as follows:

- *Actual cost matrix*: This matrix $C[i, j]$ represents the traveling time between the end of observing session i and the beginning of observing session j. The size of this matrix is dependent on the required number of sessions to survey the whole network [9].
- *The initial quantity of pheromone*: The initial quantity of pheromone τ_0, which is equivalent to the intensity of trail at time (0), should be set to an arbitrarily small value ($0 \leq \tau_0 < 1$). During the run of the program, this initial pheromone matrix is updated at each iteration and the final outcome is the general pheromone matrix $\tau[i, j]$. This general matrix provides important information about the strong and weak paths which help and direct the surveyors to select next sessions to be observed.
- *Tabu list*: The tabu list is a data structure which is correlated to each ant in order to prevent ants from observing a session more than once. In this list, the sessions already observed are memorized up to time t is forbidden to observe them again before it has completed surveying the whole network. After each iterations, the tabu list is emptied and then ants are free again to choose their ways starting from a random initial session.

The control parameters govern the workings of the GPS-ACO technique itself and are mainly concerned with pheromone information.

- *The intensity control parameter* ($\alpha \geq 0$)*:* This parameter controls the relative weight of pheromone trail intensity τ_{ij} during the selection process of the following sessions to be observed. If $\alpha = 0$ (ants do not communicate), the nearest sessions are more likely to be selected by ants. On the other hand, high values of α (ants communicate) means that the trail is very important and therefore current ants tend to choose paths previously chosen by other ants.
- *The visibility control parameter* ($\beta \geq 0$)*:* The function of parameter β is to put more or less emphasis on distance versus pheromone. This parameter has positive impact on the schedule quality and this can be seen in adjusting the relative importance of visibility ν_{ij} when evaluating the cost of a path in the schedule. If $\beta = 0$, only pheromone intensity is functioning and this will lead to the rapid selection of schedules that may not be optimal.
- *The evaporation control parameter* ($0 \leq \tau < 1$)*:* These parameter is used to adjust the magnitude of the pheromone laid by an ant when constructing its schedule and therefore is called the trail resistance.
- *The stopping criteria:* Several terminating criteria were adopted and the simplest one is to terminate the process after a pre-defined number of iterations. Each iteration of the GPS-ACO process requires two steps, to create a schedule (ant cycle); construction of a schedule and updating the level of pheromone. Another stopping criterion is to terminate the search when no further improvement can be observed for a fixed number of iterations.

5.2.1 ACO Implementation

In our implementation we use MAX-MIN Ant System (MMAS) [11], and Ant Colony System (ACS) [12], which are ones of the best ant approaches. In MMAS the main is using fixed upper bound τ_{max} and lower bound τ_{min} of the pheromone trails. Thus accumulation of big amount of pheromone by part of the possible movements and repetition of same solutions is partially prevented. The main features of MMAS are:

- Strong exploration to the space search of the best found solution. This can be achieved by only allowing one single ant to add pheromone after each iteration, the best one.
- Wide exploration of the best solution. After the first iteration the pheromone trails are reinitialized to τ_{max}. In the next iteration only the movements that belong to the best solution receive a pheromone, while other pheromone values are only evaporated.

The aim of using only one solution is to make solution elements, which frequently occur in the best found solutions, get large reinforcement. To avoid stagnation of the search, the range of possible pheromone value on each movement is limited to an interval $[\tau_{min}, \tau_{max}]$. τ_{max} is an asymptotic maximum of τ_{ij} and $\tau_{max} = 1/(1-\rho)C(V^*)$, while $\tau_{min} = 0.087\tau_{max}$. Where V^* is the optimal solution and V_{best} is the iteration best solution and $i, j = 1, \dots, n$. V^* is unknown, therefore we use V_{best} instead of V^*.

The main features of ACS are:

- Strong exploration of the space search of the best found solution.
- The pheromone of the worse solutions is decreased and partially prevents their repetition.

The pheromone corresponding to constructed by ants solutions is updated using local and global update rules. While ants build their solution at the same time they locally update the pheromone level of the visited paths. The aim is to decrease the pheromone level of used paths while this one on unused stay unchanged. Thus we partially prevent accumulation of big amount of pheromone in one side, and the pheromone level never becomes less than τ_0 in another. It is a kind of diversification of the search in the search space.

When all ants have completed their solutions, the pheromone level is updated by applying the global update rule. Only the pheromone corresponding to the best found solution is increased. The global update rule is intended to provide a greater amount of pheromone on the paths of the best solution. It is a kind of intensification of the search around the best found solution.

In both implementations we use heuristic information equals to one over the cost of the session.

5.2.2 Experimental Results

One of the main things in implementation of ACO algorithms is graph of the problem. In our implementation the nodes of the graph correspond to the stations and the arcs correspond to the cost of the sessions. The ants deposit the pheromone on the arcs.

To achieve good results we need carefully choose the ACO parameter settings and to determine structural elements of the algorithms. In every iteration ants start to construct their solution from random node of the graph, therefore the number of the ants can be much less than the number of nodes (stations). Thus we decrease needed computing resources like time and memory. Experimentally we found that 10 ants are enough. Using more ants we increase the computational time without improving achieved results, when the number of iterations is the same. Tabu list is associated with each ant in order to prevent it from visiting a node more than one.

The control parameters govern the workings of the algorithms and mainly concerned with pheromone information and transition probability. The parameter α is the pheromone intensity parameter and β is heuristic control parameter. When the value of α is high, the pheromone information is more important when the ant choose next node to move to. When the β is high the heuristic information is more important. After some tests we found that best values for transition probability control parameters for GPS surveying problem are $\alpha = 1$ and $\beta = 2$.

Parameter ρ is used in pheromone update rules to diversify the search by regulating the influence of the old pheromone. $\rho \in (0, 1)$, when ρ is close to 0 only the new added pheromone is important in the next iteration. When ρ is close to 1, the influence of old pheromone (experience from previous iterations) is great. We found that our algorithms achieve best results when $\rho = 0.2$.

The initial pheromone value τ_0 has not influence in algorithm performance. The important is τ_0 to be less than $\Delta\tau_{ij}$. Therefore we choose initial pheromone value to be $\tau_0 = 0.005$.

For realistic comparison of performance of ACS and MMAS algorithms for GPS surveying problem we use pure ACO algorithms without including any local search procedures. The number of iterations is equal to the number of the stations.

In this section we analyze the experimental results obtained using ACS and MMAS algorithms described in previous section. Like a test problems we use real data from Malta and Seychelles GPS networks. The Malta GPS network is composed of 38 sessions and the Seychelles GPS network is composed of 71 sessions. We use 6 larger test problems from http://www.informatik.uni-heidelberg.de/groups/comopt/software/TSLIB95/ATSP.html. The test problems range from 100 to 443 sessions.

For every experiment, the results are obtained by performing 30 independent runs, then averaging the fitness values obtained in order to ensure statistical confidence of the observed difference. Analysis of the data using ANOVA/Kruskal-Wallis test has been used to get statistical confidence of the level 95% of the results.

In Table 5.1 we show the achieved costs for every test problem. For comparison reason we use same parameter settings for the both algorithms. With bold are the minimal and best achieved costs.

Table 5.1 MMAS and ACS comparison

Tests	Sessions	ACS	ACS-best	MMAS	MMAS-best
Malta	38	902	900	902	900
Seychelles	71	920	853	**912**	852
rro124	100	41584	40423	**41276**	**39723**
ftv170	170	3395	3362	**3356**	**3115**
rgb323	323	1688	1680	**1665**	**1625**
rgb358	358	1734	1711	**1675**	**1646**
rgb403	403	3499	3490	**3428**	**3362**
rgb443	443	3806	3795	**3754**	**3705**

Analyzing the results we conclude that MMAS algorithm outperforms ACS algorithm. Minimal costs are obtained by MMAS for the most of the tests. For the Malta and ftv170 there are not relative difference between the two algorithms.

Lets compare our results with results in [13]. Like tests they use only small examples of Malta and Seychelles and to improve the performance of their ACO algorithm they combine it with local search procedure. They report only the best results they found as follows: 895 for Malta and 853 for Seychelles by 200 iterations. Thus we can conclude that our algorithm is better, because we achieve similar best cost for the same test problems, but without local search and with less number of iterations. Moreover we observe larger difference between the algorithms in larger test problems.

In Fig. 5.1 we plot the solution cost comparison for test problem rbg443. We observe the improvement of the results during the search processes and we can conclude that MMAS algorithm outperforms ACS for GPS surveying problem.

5.3 Hybrid ACO Algorithm for GPS Surveying Problem

Very often ACO algorithms are combined with appropriate local search procedure to improve the algorithm performance and quality of the achieved solutions.

5.3.1 Local Search Strategies

The main concept of LS is searching the local neighborhood of the current solution. The Local Search (LS) method (move-by-move method) perturbs a given solution to generate different neighborhoods using a move generation mechanism. LS attempts to improve an initial solution by a series of local improving changes. A move-generation is a transition from a solution S to another one $S' \in V(S)$ in one

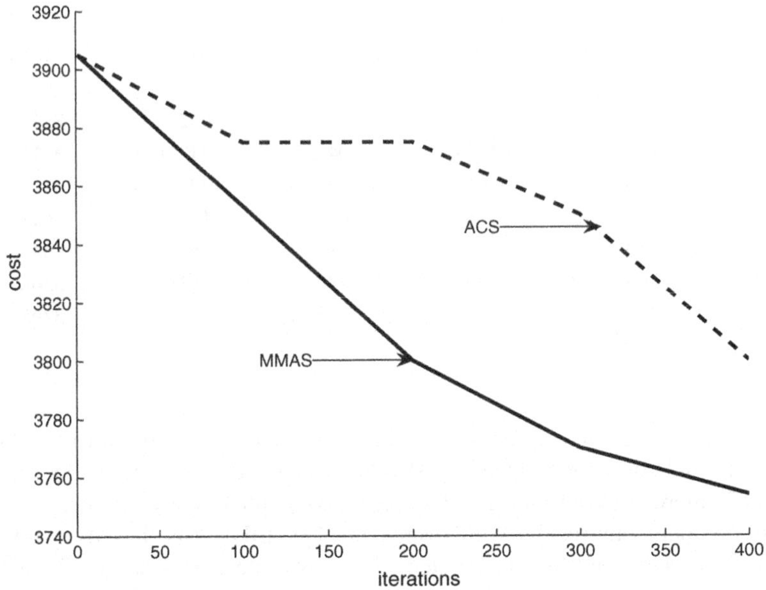

Fig. 5.1 Comparison between ACS and MMAS algorithms for rbg443 test problem

step. These solutions are selected and accepted according to some pre-defined criteria. The returned solution S' may not be optimal, but it is the best solution in its local neighborhood $V(S)$. A local optimal solution is a solution with the local minimal possible cost value. Knowledge of a solution space is the essential key to more efficient search strategies. These strategies are designed to use this prior knowledge and to overcome the complexity of an exhaustive search by organizing searches across the alternatives offered by a particular representation of the solution. The main purpose of LS implementation is to speed up and improve the solutions constructed by the ACO.

In this section, various local search structures $L(k, l)$, that satisfy the requirements of the GPS network, have been developed and implemented to search the schedule space more effectively, where k is the number of generated neighbor solutions and l is the number of used perturbation method. The problem is represented by graph. The folliwing LS procedure has been proposed:

1. Nodes Sequential Swaps: **for** $i = 1$ **to** $n - 1$; **for** $j = i + 1$ **to** n; swap a_i and a_j, [10];
2. Nodes Random Swaps: two nodes are chosen randomly and are swapped;
3. Randomly Delete an Edge: let the current solution is $(a_1, a_2, \ldots, a_i, a_{i+1}, \ldots, a_n)$. The edge $(i, i + 1)$ is randomly chosen and deleted. The new solution is $(a_{i+1}, \ldots, a_n, a_1, \ldots, a_i)$;
4. Greedy Delete an Edge: The longest (most expensive) edge is deleted. The new solution is constructed as in upper case;

5. Randomly Delete 2 Edges: Let the current solution is $(a_1, a_2, \ldots, a_i, a_{i+1}, \ldots, a_j, a_{j+1}, \ldots, a_n)$. The edges $(i, i+1)$ and $(j, j+1)$ are randomly chosen and deleted. The new solutions are $(a_{i+1}, \ldots, a_j, a_1, \ldots, a_i, a_{j+1}, \ldots, a_n)$, $(a_{j+1}, \ldots, a_n, a_{i+1}, \ldots, a_j, a_1, \ldots, a_i)$, $(a_1, \ldots, a_i, a_{j+1}, \ldots, a_n, a_{i+1}, \ldots, a_j)$;
6. Greedy Delete 2 Edges: The two longest edges are deleted. The new solutions are prepared as in the upper case.

When the ants construct their solutions we improve the current best applying LS procedures. The pheromone updating is with respect of the improved after LS solutions. We can delete more than two edges, but after several tests we established that the running time increases proportionally to the number of deleted edges without improvement of the solution.

5.3.2 Experimental Results

This section reports on the computational experience of the ACO (MMAS variant) coupled with various local search procedures. For the tests are used the same examples like in previous section. To keep low the running time, the neighbor set consists of as many solutions as the number of the sessions. The number of iterations is equal to the size of the problem.

In Table 5.2 we show achieved costs for all tested problems, with bold are the minimal achieved costs. After analysis of the results, we conclude that LS procedure $L(n, 5)$ outperforms others. Comparing MMAS combined with different LS procedures the best results are obtained by $L(n, 5)$ for most of the tests except the ftv170. For this test problem there is not relative difference between $L(n, 3)$ and $L(n, 5)$. We observe that there are not relative difference between achieved costs by $L(n, 1)$, $L(n, 2)$ and achieved costs without LS. We conclude that $L(n, 1)$ and $L(n, 2)$ do not improve achieved by MMAS solutions.

Table 5.2 MMAS algorithm plus local search procedures

Tests	LS sessions	L(n,1)	L(n,2)	L(n,3)	L(n,4)	L(n,5)	L(n,6)
Malta	38	902	902	895	895	**872**	902
Seychelles	71	920	920	915	915	**851**	893
rro124	100	41276	41276	41004	41004	**39971**	41281
ftv170	170	3356	3356	**3229**	3290	**3266**	3320
rgb323	323	1661	1661	1691	1666	**1378**	1423
rgb358	358	1680	1680	1689	1702	**1477**	1549
rgb403	403	3428	3428	3401	3401	**2408**	2710
rgb443	443	3751	3751	3838	3838	**2631**	3349

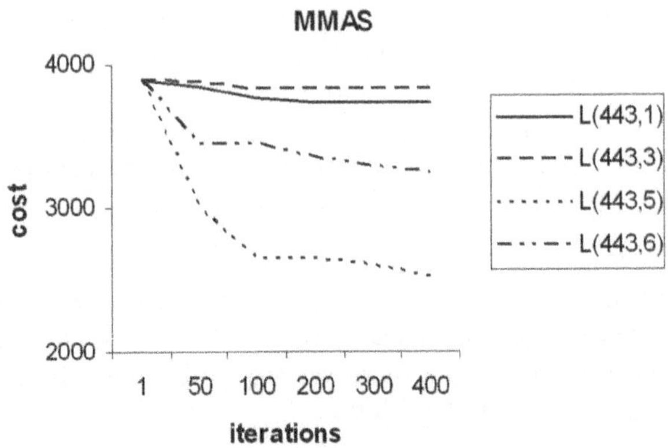

Fig. 5.2 MMAS with local search procedures for rbg443 test problem

In Fig. 5.2 we plot the solution cost comparison for test problem rbg443 applying $L(443, 1)$, $L(443, 3)$, $L(443, 5)$ and $L(443, 6)$. We observe that $L(443, 5)$ and $L(443, 6)$ achieve better results than others. The best cost is achieved by $L(443, 5)$. The costs achieved by $L(443, 1)$ are statistically similar to the costs achieved without LS, thus on Fig. 5.2 we observe the improvement of the results using LS. Local search procedure $L(n, 5)$ performs better than $L(n, 6)$, because random deletion leads to large diversification.

5.4 ACO with Environment Changes

We propose a variant on the well-known Ant Colony Optimization (ACO) general framework where we introduce the environment to play an important role during the optimization process. Together with diversification and intensification, the environment is introduced with the aim of avoiding the search to get stuck at local optima. In this work, the environment is simulated by means of the Logistic map, that is used in ACO for perturbing the update of the pheromone trails. We verify the effectiveness of this new idea by solving instances of the GPS Surveying Problem (GSP), which is known to be NP-hard. Preliminary experiments show that our environment ACO (*e*ACO), with variable environment, outperforms the standard ACO on the entire set of GSP instances that we consider.

5.4.1 Simulating the Environment

The Logistic map is a quadratic dynamical equation proposed in 1938 as a demographic model [14]. It is a rather simple quadratic polynomial:

$$x_{n+1} = r x_n (1 - x_n), \qquad n > 0, \tag{5.1}$$

where x_n represents the population size at time n and r is a constant, named growth coefficient. Given $x_0 \in [0, 1]$ and a value for $r \in [0, 4]$, this dynamical equation can either converge or be chaotic. In the first case, given any $x_0 \in [0, 1]$, x_n tends to the so-called "attraction domain". In the second case, x_n never converges, but it can rather take, in an apparent random way, any possible value in the range $[0, 1]$.

Figure 5.3 shows the behavior of the Logistic map for different values of r in the range $[2, 4]$ (its behavior is linear in the range $[0, 2]$). On the x-axis, we consider a discrete subset of 3000 equidistant values for r between 2 and 4; on the y-axis, for every considered value for r, we report the corresponding attraction domain. In order to identify the attraction domains, we take 1500 equidistant points in the interval $[0, 1]$ and we apply Eq. (5.1) 1000 times for each of them. For small values of r, the Logistic map always converges to one single point, i.e. the attraction domain consists of one point only. The first bifurcation appears when $r = 3$, where the attraction domain consists of 2 points; then there is another bifurcation when $r = 1 + \sqrt{6}$, so that the attraction domain consists of 4 points. For larger values for r, the Logistic map experiences other bifurcations, and it can be chaotic for some subintervals of r. However, in these chaotic regions, it is still possible to identify regular attraction domains. For example, in Fig. 5.4, the same graphic reported in Fig. 5.3 is zoomed in the region $r = [3.901, 3.908]$, where this phenomenon is clearly shown. Regular regions, that can be glimpsed in Fig. 5.3, still contain bifurcations. Moreover, we can notice that the whole graphic in Fig. 5.3 reappears in our zoomed region. Other regular attraction domains can be identified by looking at tighter subintervals of r, as well as other copies of the entire graphic. The graphic in Fig. 5.3 is in fact a fractal, because of its self-similarity [15, 16].

We simulate regular and chaotic changes of environment in ACO by introducing the Logistic map in Eq. (5.1), which is used in ACO for updating the pheromone trails. In the hypothesis the objective function of the considered problem is positive

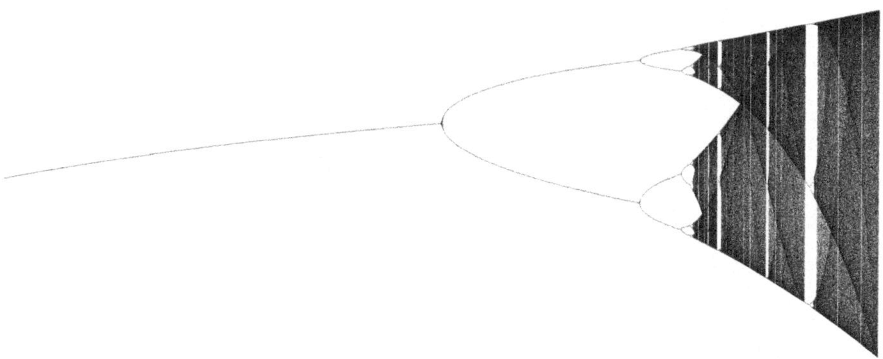

Fig. 5.3 The behavior of the Logistic map for different values of r (on the x-axis, for $r = 2$ to 4 in the figure). The attraction domain can be either regular or chaotic

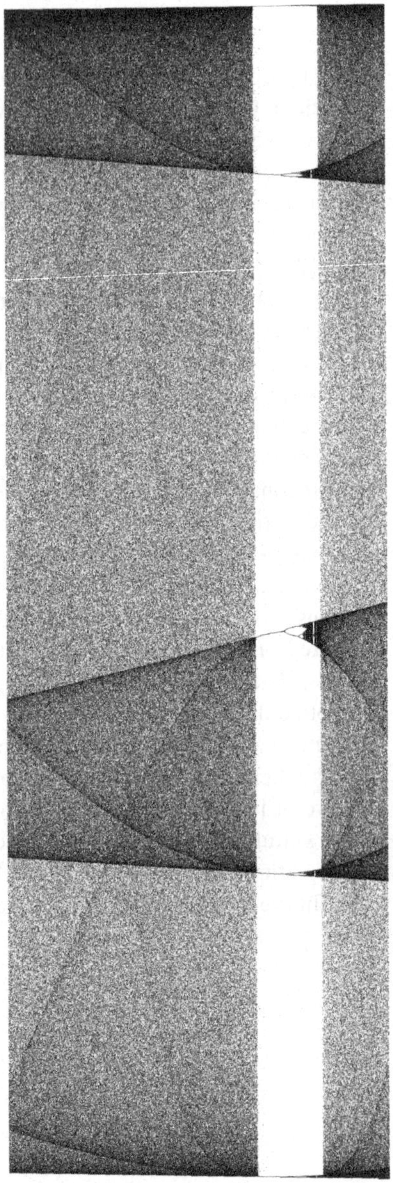

Fig. 5.4 The behavior of the Logistic map for values of r in the interval [3.901, 3.908]. This region of the Logistic map is generally chaotic, but regular attraction domains (with the typical bifurcations) can still be identified

and greater than 1, the term $1/f(X)$ in Eq. (5.2) has always values ranging between 0 and 1. It can therefore take the place of x_0 in the Logistic map, so that a perturbed value x_1 can be computed, for a given value of $r \in [0, 4]$. The equation for updating the pheromone therefore becomes:

$$\tau_{uv} = \tau_{uv} + r \cdot \frac{1}{f(X)} \cdot \left(1 - \frac{1}{f(X)}\right). \tag{5.2}$$

With this simple change in the rule for updating the pheromone, we artificially perturb the environment of the ants, which would otherwise only depend on the solution fitness values. Different values for r can produce different environment changes, depending on the behavior of the Logistic map. For values of r for which the Logistic map converges, the pheromone levels added to τ_{uv} tend to be constant, reducing in this way the effects of good-quality solutions, that might mislead the ants towards a local optimum. For values of r for which the Logistic map behaves instead chaotically, the environment is dominant on the choices of the ants, as the pheromone update mostly depend on the simulated environment, rather than on the actual fitness value.

We refer to ACO with environment changes as *environmental* ACO (*e*ACO). In this work, we present some preliminary experiments where *e*ACO is employed for solving the GSP.

5.4.2 Computational Experiments

As test cases, we consider data from two real networks: Malta [13], composed by 38 sessions, and Seychelles [17], composed by 71 sessions. We also consider larger instances designed for testing the ATSP, which are freely available on the Internet.[1]

In our *e*ACO implementation the pheromone update is performed by applying Eq. (5.2). Finally, the values for the transition probability parameters α and β are fixed to 1 and 2, respectively. These values were identified as the optimal ones for ACO when solving instances of the GSP [4]. In the following experiments, we will focus on the quality of the found solutions, rather than on the algorithms' performances. In fact, the increase in complexity for using Eq. (5.2), can be neglected when considering the overall algorithms' complexity.

Table 5.3 shows some computational experiments for different values of r. Average values over 30 runs are reported in the table. *e*ACO is able to identify better quality solutions in all experiments and for almost all used values for r. This shows that, in fact, a variable environment for the ants, instead of a constant one, gives benefits to the search. For values of r equal to 1, 2 and 3, the Logistic map converges to one unique value; it is instead chaotic for $r = 4$. It seems therefore that the best results can be achieved when the environment tends to homogenize the pheromone

[1] http://www.informatik.uni-heidelberg.de/groups/comopt/software/TSLIB95/ATSP.html.

Table 5.3 Comparison between ACO and *e*ACO on a set of GSP instances

		ACO	*e*ACO			
Instances	$\|V\|$		$r = 1$	$r = 2$	$r = 3$	$r = 4$
Malta	38	899.50	**897.00**	**897.00**	**897.00**	900.33
Seyshels	71	922.06	905.73	905.60	**887.33**	906.73
kro124p	100	40910.60	**40725.40**	40799.30	40753.00	40803.76
ftv170	171	3341.93	3314.20	**3313.76**	3319.83	3338.53
rgb323	323	1665.90	1654.40	**1648.66**	1649.43	1649.53
rgb358	358	1692.66	**1679.63**	1689.00	1682.80	1685.95
rgb403	403	3428.56	3413.63	3392.23	3393.76	**3386.10**
rgb443	443	3765.80	3749.93	3742.86	**3742.43**	3754.50

trails. Notice, however, that only one experiment with $r = 4$, when the environment is chaotic, was not able to provide a better solution (w.r.t. the one provided by the standard ACO).

5.5 InterCriteria Analysis of ACO Application to GPS Surveying Problems

In this chapter InterCriteria Analysis (ICrA) has been applied for analysis of an Ant Colony Optimization (ACO) algorithm used to provide near-optimal solutions for Global Positioning System surveying problem.

5.5.1 InterCriteria Analysis

Following [18, 19], an Intuitionistic Fuzzy Pair (IFP) [20] with the degrees of "agreement" and "disagreement" between two criteria applied on different objects is obtained. An IFP is an ordered pair of real non-negative numbers $\langle a, b \rangle$ such that: $a + b \leq 1$.

Let an Index Matrix (IM) (see [21]) whose index sets consist of the names of the criteria (for rows) and objects (for columns) be given. The elements of this IM are further supposed to be real numbers. An IM with index sets consisting of the names of the criteria (for rows and for columns) with elements IFPs corresponding to the "agreement" and "disagreement" of the respective criteria will be obtained. Two things are further assumed: (i) all criteria provide an evaluation for all objects and all these evaluations are available; (ii) all the evaluations of a given criteria can be compared amongst themselves.

The set of all objects being evaluated is denoted by O, and the set of values assigned by a given criteria C to the objects by $C(O)$, i.e.

$$O \stackrel{\text{def}}{=} \{O_1, O_2, \ldots, O_n\}, \;\; C(O) \stackrel{\text{def}}{=} \{C(O_1), C(O_2), \ldots, C(O_n)\}.$$

Let $C^*(O) \stackrel{\text{def}}{=} \{\langle x, y\rangle | \, x \neq y \;\&\; \langle x, y\rangle \in C(O) \times C(O)\}$.

In order to compare two criteria, the vector of all internal comparisons of each criteria which fulfill exactly one of three relations R, $\overline{R}$ and $\tilde{R}$ must be constructed. It is required that for a fixed criterion C and any ordered pair $\langle x, y\rangle \in C^*(O)$ it is true:

$$\langle x, y\rangle \in R \Leftrightarrow \langle y, x\rangle \in \overline{R} \tag{5.3}$$

$$\langle x, y\rangle \in \tilde{R} \Leftrightarrow \langle x, y\rangle \notin (R \cup \overline{R}), \tag{5.4}$$

$$R \cup \overline{R} \cup \tilde{R} = C^*(O). \tag{5.5}$$

From the above it is seen that only a subset of $C(O) \times C(O)$ has to be considered for the effective calculation of the vector of internal comparisons (denoted further by $V(C)$), since from 5.3 to 5.5 it follows that if the relation between x and y is known, the relation between y and x is known as well. Thus, only lexicographically ordered pairs $\langle x, y\rangle$ are considered. Let, for brevity, $C_{i,j} = \langle C(O_i), C(O_j)\rangle$. Then, for a fixed criterion C the following vector is constructed:

$$V(C) = \{C_{1,2}, C_{1,3}, \ldots, C_{1,n}, C_{2,3}, C_{2,4}, \ldots, C_{2,n}, C_{3,4}, \ldots, C_{3,n}, \ldots, C_{n-1,n}\}.$$

It can be easily seen that it has exactly $\frac{n(n-1)}{2}$ elements. Further, to simplify our considerations, the vector $V(C)$ is replaced with $\hat{V}(C)$, where for each $1 \leq k \leq \frac{n(n-1)}{2}$ for the k-th component it is true:

$$\hat{V}_k(C) = \begin{cases} 1 & \text{iff } V_k(C) \in R, \\ -1 & \text{iff } V_k(C) \in \overline{R}, \\ 0 & \text{otherwise} \end{cases} \tag{5.6}$$

Then, when comparing two criteria, the degree of "agreement" between the two is the number of matching components (divided by the length of the vector for normalization purposes). The degree of "disagreement" is the number of components of opposing signs in the two vectors (again normalized by the length).

The above described algorithm for calculating the degrees of "agreement" (μ) and degrees of "disagreement" (ν) between two criteria C and C' is realized in Matlab environment according to [22].

5.5.2 ICrA Results

In Table 5.4 the results of every 30 runs for the eight GPS Surveying Problems (GSPs) are presented. The average results of the first 5 runs, first 10 runs, etc, and finaly of all 30 runs are calculated and presented in Table 5.5. ICA has been applied on these results.

Table 5.4 Results of 30 runs

	GSP_1	GSP_2	GSP_3	GSP_4	GSP_5	GSP_6	GSP_7	GSP_8
run1	905	897	41,773	3333	1670	1711	3434	3786
run2	900	914	40,486	3115	1636	1648	3412	3791
run3	900	934	41,005	3279	1663	1668	3454	3731
run4	895	910	41,311	3230	1652	1657	3416	3754
run5	895	927	42,107	3267	1660	1684	3384	3729
run6	895	908	40,089	3422	1641	1697	3397	3723
run7	900	963	41,384	3299	1665	1699	3486	3723
run8	895	915	40,894	3448	1679	1702	3454	3779
run9	900	865	40,471	3314	1681	1688	3396	3769
run10	895	923	41,004	3326	1661	1681	3474	3772
run11	900	911	40,776	3450	1692	1695	3447	3776
run12	895	918	41,299	3408	1615	1708	3484	3784
run13	900	959	40,541	3265	1691	1689	3435	3720
run14	900	930	41,046	3391	1661	1715	3382	3794
run15	900	963	40,693	3358	1692	1719	3442	3750
run16	895	934	42,219	3405	1679	1664	3364	3733
run17	900	918	40,947	3290	1646	1690	3422	3717
run18	900	914	40,602	3398	1654	1673	3395	3803
run19	895	930	40,827	3630	1677	1684	3431	3827
run20	905	963	39,244	3263	1684	1703	3428	3749
run21	920	916	41,137	3341	1694	1726	3464	3701
run22	900	914	41,773	3370	1659	1693	3430	3778
run23	900	916	39,477	3291	1646	1675	3422	3754
run24	900	963	39,096	3338	1678	1735	3446	3758
run25	895	913	40,604	3409	1698	1652	3437	3819
run26	900	964	41,147	3302	1642	1699	3393	3791
run27	900	865	40,984	3286	1641	1692	3407	3785
run28	905	897	41,311	3333	1657	1705	3472	3803
run29	895	928	41,747	3335	1674	1705	3405	3796
run30	900	890	41,324	3362	1689	1723	3444	3779

Table 5.5 IM for ICrA

	C_1	C_2	C_3	C_4	C_5	C_6
GSP_1	899.00	898.00	898.33	898.50	899.40	899.50
GSP_2	916.40	915.60	922.47	924.80	924.72	922.07
GSP_3	41,336.40	41,052.40	40,991.93	40,935.90	40,832.20	40,910.60
GSP_4	3244.80	3303.30	3327.00	3344.55	3345.60	3341.93
GSP_5	1656.20	1660.80	1663.93	1664.95	1666.96	1665.90
GSP_6	1673.60	1683.50	1690.73	1688.75	1690.24	1692.67
GSP_7	3420.00	3430.70	3433.13	3426.85	3429.44	3428.57
GSP_8	3758.20	3755.70	3758.73	3760.50	3760.80	3765.80

Resulting degrees of "agreement" ($\mu_{C,C'}$) are as follows:

$$
\mathbf{IM}_1 = \begin{array}{c|cccccccc}
 & GSP_1 & GSP_2 & GSP_3 & GSP_4 & GSP_5 & GSP_6 & GSP_7 & GSP_8 \\
\hline
GSP_1 & \mathbf{1} & 0.60 & 0.27 & 0.67 & 0.73 & 0.67 & 0.33 & 0.87 \\
GSP_2 & 0.60 & \mathbf{1} & 0.27 & 0.80 & 0.73 & 0.53 & 0.47 & 0.73 \\
GSP_3 & 0.27 & 0.27 & \mathbf{1} & 0.07 & 0 & 0.20 & 0.40 & 0.13 \\
GSP_4 & 0.67 & 0.80 & 0.07 & \mathbf{1} & 0.93 & 0.73 & 0.53 & 0.80 \\
GSP_5 & 0.73 & 0.73 & 0 & 0.93 & \mathbf{1} & 0.80 & 0.60 & 0.87 \\
GSP_6 & 0.67 & 0.53 & 0.20 & 0.73 & 0.80 & \mathbf{1} & 0.67 & 0.80 \\
GSP_7 & 0.33 & 0.47 & 0.40 & 0.53 & 0.60 & 0.67 & \mathbf{1} & 0.47 \\
GSP_8 & 0.87 & 0.73 & 0.13 & 0.80 & 0.87 & 0.80 & 0.47 & \mathbf{1}
\end{array} \tag{5.7}
$$

Resulting degrees of "disagreement" ($\nu_{C,C'}$) are as follows:

$$
\mathbf{IM}_2 = \begin{array}{c|cccccccc}
 & GSP_1 & GSP_2 & GSP_3 & GSP_4 & GSP_5 & GSP_6 & GSP_7 & GSP_8 \\
\hline
GSP_1 & \mathbf{0} & 0.40 & 0.73 & 0.33 & 0.27 & 0.33 & 0.67 & 0.13 \\
GSP_2 & 0.40 & \mathbf{0} & 0.73 & 0.20 & 0.27 & 0.47 & 0.53 & 0.27 \\
GSP_3 & 0.73 & 0.73 & \mathbf{0} & 0.93 & 1 & 0.80 & 0.60 & 0.87 \\
GSP_4 & 0.33 & 0.20 & 0.93 & \mathbf{0} & 0.07 & 0.27 & 0.47 & 0.20 \\
GSP_5 & 0.27 & 0.27 & 1 & 0.07 & \mathbf{0} & 0.20 & 0.40 & 0.13 \\
GSP_6 & 0.33 & 0.47 & 0.80 & 0.27 & 0.20 & \mathbf{0} & 0.33 & 0.20 \\
GSP_7 & 0.67 & 0.53 & 0.60 & 0.47 & 0.40 & 0.33 & \mathbf{0} & 0.53 \\
GSP_8 & 0.13 & 0.27 & 0.87 & 0.20 & 0.13 & 0.20 & 0.53 & \mathbf{0}
\end{array} \tag{5.8}
$$

Regarding InterCriteria analysis we obser that the ACO algorithm performs in similar way for the GSP_2, GSP_4, GSP_5 and GSP_8. They are GPS networcs with different number f sessions, but may be these networks have similar structure, therefore the value of "agreement" is high. For other networks we can conclud that they have very different structure. The GSP that are in positive consonance are shown in Fig. 5.5. For these GSP ACO obtains close and similar results, i.e. algorithm per-

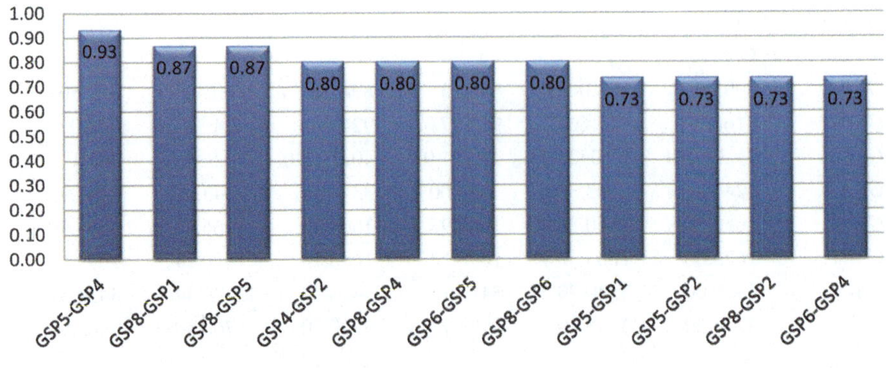

Fig. 5.5 GSP relation in positive consonance scale

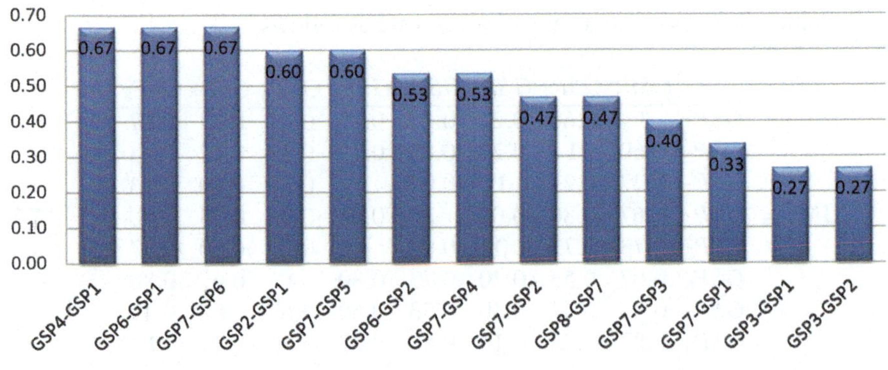

Fig. 5.6 GSP relation in dissonance scale

formance is identical solving the given tasks. In Fig. 5.6 are shown these couples of GSP for which ACO does not have identical performance, i.e. these GSP are in dissonance. Finally, in Fig. 5.7 are presented couples of GSP which are in negative consonance. The definition of the positive/negative consonance and dissonance are according [23].

5.6 Conclusion

The GPS surveying problem is addressed in this paper. Instances containing from 38 to 443 sessions have been solved using MMAS and ACS algorithms. For both algorithms we use same parameter settings. A comparison of the performance of the both ACO algorithms applied on various GPS networks is reported. The MMAS algorithm outperforms ACS algorithm. The obtained results are encouraging and the ability of the developed techniques to generate rapidly high-quality solutions for observing GPS networks can be seen. The problem is important because it arises in

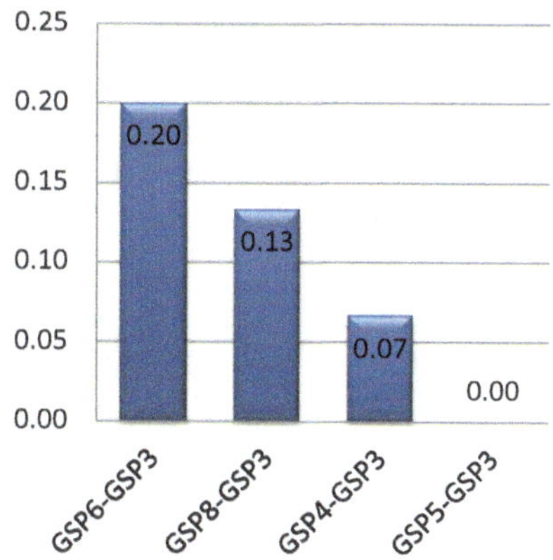

Fig. 5.7 GSP relation in negative consonance scale

wireless communications like GPS and mobile phone and can improve the services in the networks. Thus the problem has an economical importance. ICrA has been applied on the obtained results. By ICrA we can verify the correctnes of the algorithm performance. It shows whether the construction of the different GPS network are similar.

References

1. Fidanova, S., Saleh, H. A.: Ant colony optimization for scheduling the surveying activities of satellite positioning networks. International Conference on Information Systems and Data Grids, Sofia Bulgaria, pp. 43–54 (2005)
2. Fidanova, S.: Hybrid heuristics algorithm for GPS surveying problem. Numer. Methods Appl., Lect. Notes Comput. Sci. **4310**, 239–248. Springer (2007)
3. Fidanova, S.: MMAS and ACS for GPS surveying problem. In: Proceeding of International Conference on Evolutionary Computing, pp. 87–91, Sofia, Bulgaria (2008)
4. Fidanova, S., Alba, E., Molina, G.: Hybrid ACO algorithm for the GPS surveying problem. Large Scale Sci. Comput., Lect. Notes Comput. Sci. **5910**, 318–325. Springer (2010)
5. Fidanova, S., Mucherino, A.: Ant Colony Optimization with Environment Changes: An Application to GPS Surveying. FedCSIS'2015, EEE Xplorer, IEEE catalog number CFP1585N-ART, pp. 495–500 (2015)
6. Fidanova, S., Roeva, O.: InterCriteria analysis of ant colony optimization application to GPS surveying problems. Issues Intuitionistic Fuzzy Sets Gen. Nets **12**, 20–38 (2016)
7. Fidanova, S., Roeva, O., Mucherino, A., Kapanova, K.: InterCriteria analysis of ANT algorithm with environment change for GPS surveying problem. Issues Intuitionistic Fuzzy Sets Gen. Nets (Dachev, Ch., Agre, G., eds.), Lecture Notes in Artificial Intelligence **9883**, 271–278. Springer (2016)
8. Fidanova, S., Roeva, O., Atanasova, V.: Ant colony optimization application to GPS surveying problems: InterCriteria analysis. Intuitionistic Fuzzy Sets Gen. Nets, Adv. Intell. Syst. Comput. **559**, 251–264. Springer (2018)
9. Dare, P.J., Saleh, H.A.: GPS network design: logistics solution using optimal and near-optimal methods. J. Geod. **74**, 467–478 (2000)
10. Saleh, H.A., Dare, P.: Effective heuristics for the GPS survey network of Malta: simulated annealing and Tabu search techniques. J. Heuristics **7**(6), 533–549 (2001)

11. Shmygelska, A., Hoos, H.H.: An ant colony optimization algorithm for the 2D and 3D hydrophobic polar protein folding problem. BMC Bioinform. **6**(30). https://doi.org/10.1186/1471-2105-6-30 (2005)
12. Dorigo, M., Gambardella, L.M.: Ant colony system: a cooperative learning approach to the traveling salesman problem. IEEE Trans. Evol. Comput. **1**, 53–66 (1997)
13. Saleh, H.A.: Ants can successfully design GPS surveying networks. GPS World **9**, 48–60 (2002)
14. Verhulst, P.F.: A note on the law of population growth. Correspondence Mathematiques et Physiques **10**, 113–121 (1938). (in French)
15. Falconer, K.: Fractal Geometry: Mathematical Foundations and Applications. Wiley, 400 pages (2013)
16. Rani, M., Agarwal, R.: Generation of fractals from complex logistic map. Chaos, Solitons and Fractals **42**, 447–452 (2009)
17. Saleh, H.A., Dare, P.J.: Heuristic methods for designing global positioning system surveying network in the Republic of Seychelles. Arab. J. Sci. Eng. **26**, 73–93 (2002)
18. Atanassov, K., Mavrov, D., Atanassova, V.: Intercriteria decision making: a new approach for multicriteria decision making, based on index matrices and intuitionistic fuzzy sets. Issues in IFSs and GNs **11**, 1–8 (2014)
19. Atanassov, K.: On Intuitionistic Fuzzy Sets Theory. Springer, Berlin (2012)
20. Atanassov, K., Szmidt, E., Kacprzyk, J.: On intuitionistic fuzzy pairs. Notes on IFS **19**(3), 1–13 (2013)
21. Atanassov, K.: On index matrices, part 1: standard cases. Adv. Stud. Contemp. Math. **20**(2), 291–302 (2010)
22. Roeva, O., Vassilev, P.: InterCriteria analysis of generation gap influence on genetic algorithms performance. Adv. Intell. Syst. Comput. **401**, 301–313 (2016)
23. Atanassov, K., Atanassova, V., Gluhchev, G.: InterCriteria analysis: ideas and problems. Notes on Intuitionistic Fuzzy Sets **21**(1), 81–88 (2015)

Chapter 6
ACO for *E. coli* Cultivation Model

The text of this chapter is based on [1–6]. A lot of proteins are produced by the modified genetically microorganisms. One of the most used host organisms in the process is the *Escherichia coli* [7]. Furthermore, the *E. coli* is still the most important host organism for the recombinant protein production. In many cases, cultivation of recombinant micro-organisms e.g. the *E. coli*, is the only economical way to produce pharmaceutical biochemicals such as: interleukins, insulin, interferons, enzymes and growth factors, etc. Simple bacteria, like the *E. coli*, are manipulated to produce these chemicals so that they are easily harvested in vast quantities for use in medicine. Scientists may know more about the *E. coli* than they do know about any other species on earth. Research on the *E. coli* accelerated after 1997, after publication of its entire genome. The scientists were able to survey all 4,288 of its genes, discovering how groups of them worked together to break down food, make new copies of the DNA and do other tasks. However, despite decades of research, there rest a lot more to know about the *E. coli*. In 2002, they formed the *International E-coli Alliance*, for organization of projects that many laboratories could work together. As knowledge of the *E. coli* grows, scientists are starting to build models of the microbe that capture some of its behavior. It is important to be able to simulate how fast the microbe will grow on various sources of food, and how its growth changes if individual genes are knocked out. These questions are best answered by application of mathematical modeling.

Modeling of biotechnological processes is a common tool in process technology. It is obvious that the model is always a simplification of the reality. This is especially true when trying to model natural systems containing living organisms. However, for many industrial relevant processes, detailed models are not available due to the insufficient understanding of the underlying phenomena. These models can be too complicated and/or impossible to be solved. Therefore the specialists try to separate the most important components, and to create simplified models, which are as close as possible to the real processes. The mathematical models are very useful and effective tools in describing those effects. They are of great importance for control, optimization, or for understanding of the process. Thus the numerical solution of

S. Fidanova, *Ant Colony Optimization and Applications*, Studies in Computational Intelligence 947, https://doi.org/10.1007/978-3-030-67380-2_6

the models is fundamental for the development of powerful, though economical, methods in the fields of bioprocess design, plant design, scale-up, optimization and bioprocess control [8, 9]. Some of the recent researches and developed models of the *E. coli* were presented in [10–15].

A common approach to model cellular dynamics is by systems of nonlinear differential equations. Obviously, parameter identification of a nonlinear dynamic model is more difficult than the linear one, as no general analytic results exist. The difficulties that may arise are such as: convergence to local solutions if standard local methods are used, over-determined models, badly scaled model function, etc. The problem is NP-hard and it is unpractical to be solved with exact or traditional numerical method. Therefore, existing research results indicate that the most useful solution method is by application of some metaheuristics. During the last decade metaheuristic techniques have been applied in a variety of areas. Heuristics can obtain suboptimal solution in ordinary situations and optimal solution in particular cases. Since the considered problem has been known to be NP-complete, using heuristic techniques can solve this problem more efficiently. Three best known (and most studied) heuristic approaches are: the iterative improvement algorithms, the probabilistic optimization algorithms, and the constructive heuristics. In this context, the evolutionary algorithms like: (a) Genetic Algorithms (GA) [16–18], (b) Evolution Strategies, (c) Ant Colony Optimization (ACO) [19–22], (d) Particle Swarm Optimization [23], (e) Tabu Search (TS) [24], (f) Simulated Annealing (SA) [25], (g) estimation of distribution algorithms, (h) scatter search, (i) path relinking, (j) greedy randomized adaptive search procedure, (k) multi-start and iterated local search, (l) guided local search, and (m) variable neighborhood search are—among others—often listed as examples of classical metaheuristics [26–28].

Obviously, they all have individual historical backgrounds and follow different paradigms and philosophies [29]. In this work the ACO is chosen as the most common direct methods used for the global optimization.

The ACO is a rapidly growing research area of population-based metaheuristics that can be used to find approximate solutions to difficult optimization problems. It is applicable for a broad range of optimization problems, can be used in dynamic applications (adapts to changes such as new distances, etc.) and in some complex biological problems [30–32]. Recall that the ACO can compete with other global optimization techniques like GAs and SA. Overall, the ACO algorithms have been inspired by the real-world ant behavior. In nature, ants usually wander randomly, and upon finding food return to their nest while laying down pheromone trails. If other ants find such a path, they are likely to not continue traveling at random, but to follow the trail instead, returning and reinforcing it (if they eventually find food). However, as time passes, the pheromone starts to evaporate. Therefore, the more time it takes for an ant to travel down the path and back again, the more time the pheromone has to evaporate and the path becomes less noticeable. A shorter path, in comparison, will be visited by more ants and thus the pheromone density remains high for a longer time. The ACO is usually implemented as a team of intelligent agents which simulate the ants behavior, walking around the graph representing the problem to solve using mechanisms of cooperation and adaptation.

In this paper ACO is applied for parameter identification of a system of the *E. coli* fed-batch cultivation process, described in terms of a mathematical model. Specifically, a system of nonlinear ordinary differential equations is proposed to model the *E. coli* biomass growth and substrate (glucose) utilization. The parameter optimization is performed using real experimental data set from the *E. coli* MC4110 fed-batch cultivation process. The cultivation was performed in the *Institute of Technical Chemistry, of the University of Hannover, Germany* during the collaboration work with the *Institute of Biophysics and Biomedical Engineering, BAS, Bulgaria*, and was funded by a grant *DFG*. The experimental data set includes records for the substrate feeding rate, concentration of biomass and substrate (glucose), and the cultivation time. In the nonlinear mathematical model considered here, the parameters that should be estimated are the maximum specific growth rate (μ_{max}), the saturation constant (k_S), and the yield coefficient ($Y_{S/X}$).

The parameter estimation is performed based upon the use of a modified Hausdorff Metric [33] and the most commonly used metric—the Least Square Regression. The Hausdorff Metrics are used in the geometric settings for measuring the distance between sets of points. They have been used extensively in areas such as computer vision, pattern recognition and computational chemistry [34–37]. The modified Hausdorff Distance is proposed to evaluate the mismatch between the experimental and the model predicted data. The results from both metrics are compared and analyzed.

6.1 Problem Formulation

The costs of developing mathematical models for the bioprocess improvement are often too high and the benefits too low. The main reason for this is related to the intrinsic complexity and non-linearity of biological systems. In general, mathematical descriptions of growth kinetics assume extensive simplifications. These models are often not accurate enough to correctly describe the underlying mechanisms. Another critical issue is related to the nature of the bioprocess models. Quite often, the parameters involved are not identifiable. Additionally, from the practical point of view, such identification would require data from specific experiments, which are themselves difficult to design and to realize. However, the estimation of model parameters with high parameter accuracy is essential for successful model development.

The real parameter optimization of simulation models, has become a research field of great interest in recent years. Nevertheless, after all completed research, this task still represents a very difficult problem. This mathematical problem, the so-called inverse problem, is a big challenge for the traditional optimization methods. In this case only the direct optimization strategies can be applied, because they exclusively use information about values of the goal function. Additional information about the goal function, like its gradients, etc., which could be used to accelerate the optimization process, is not available. Since an evolution of a goal for one string is provided by one simulation run, completing of the optimization algorithm may require a lot of computation time. Therefore, various metaheuristics are used as an alternative to surmount the parameter estimation difficulties.

6.1.1 Problem Model

The general state space dynamical model of the process of interest was described by Bastin and Dochain in [38]. It is accepted as representing the dynamics of an n components and m reactions bioprocess:

$$\frac{dx}{dt} = K\varphi(x,t) - Dx + F - Q. \tag{6.1}$$

Here, x is a vector representing the state components; K is the yield coefficient matrix; φ is the growth rates vector; the vectors F and Q are the feed rates and the gaseous outflow rates. The scalar D is the dilution rate, which will be the manipulated variable, and which is defined as follows:

$$D = \frac{F_{in}}{V} \tag{6.2}$$

where F_{in} is the influent flow rate and V is the bioreactor volume.

Application of the general state space dynamical model [38] to the *E. coli* cultivation fed-batch process leads to the following nonlinear differential equation system [39]:

$$\frac{dX}{dt} = \mu_{max}\frac{S}{k_S + S}X - \frac{F_{in}}{V}X \tag{6.3}$$

$$\frac{dS}{dt} = -\frac{1}{Y_{S/X}}\mu_{max}\frac{S}{k_S + S}X + \frac{F_{in}}{V}(S_{in} - S) \tag{6.4}$$

$$\frac{dV}{dt} = F_{in} \tag{6.5}$$

where:

X	– biomass concentration, $[g/l]$;
S	– substrate concentration, $[g/l]$;
F_{in}	– feeding rate, $[l/h]$;
V	– bioreactor volume, $[l]$;
S_{in}	– substrate concentration in the feeding solution, $[g/l]$;
μ_{max}	– maximum value of the specific growth rate, $[h^{-1}]$;
k_S	– saturation constant, $[g/l]$;
$Y_{S/X}$	– yield coefficient, $[-]$.

The mathematical formulation of the nonlinear dynamic model Eqs. (6.3)–(6.5) of the *E. coli* fed-batch cultivation process is described according to the mass balance and the model is based on the following a'priori assumptions:

- the bioreactor is completely mixed;
- the main products are biomass, water and, under some conditions, acetate;
- the substrate glucose is consumed mainly oxidatively and its consumption can be described by the Monod kinetics;
- the variation in the growth rate and the substrate consumption do not significantly change the elemental composition of the biomass, thus only balanced growth conditions are assumed;
- parameters, e.g. temperature, pH, or pO_2, are controlled at their individual constant set points.

For the parameter estimation problem the real experimental data of the *E. coli MC4110* fed-batch cultivation process is used. Off-line measurements of the biomass and on-line measurements of the glucose concentration are used in the identification procedure. The cultivation condition and the experimental data have been published in [40]. Here only the fermentation conditions are described.

The fed-batch cultivation of the *E. coli* MC4110 is performed in a 2l bioreactor (Bioengineering, Switzerland), using a mineral medium [41], in the *Institute of Technical Chemistry, University of Hannover*. Before inoculation, a glucose concentration of 2.5 g/l is established in the medium. Glucose in the feeding solution is 100 g/l. The initial liquid volume is 1350 ml. The pH is controlled at 6.8 and the temperature is kept constant at 35 °C. The aeration rate is kept at 275 l/h air, the stirrer speed at start 900 rpm, and after 11 hours the stirrer speed is increased in steps of 100 rpm. At end the stirrer sped reaches 1500 rpm. Oxygen is controlled at around 35%.

Off-line analysis
For the off-line glucose measurements, as well as the biomass and the acetate concentration determination, samples of about 10 ml are taken approximately at every hour. Off-line measurements are performed by using the Yellow Springs Analyser (Yellow Springs Instruments, USA).

On-line analysis
For the on-line glucose determination a flow injection analysis (FIA) system has been employed, using two pumps (ACCU FM40, SciLog, USA) for the continuous sample and the carrier flow rate. To reduce the measurement noise the continuous-discrete extended Kalman filter was used [41].

Glucose measurement and control system
For on-line glucose determination, the same FIA system has been employed for the continuous sample and the carrier flow rate at 0.5 ml/min and 1.7 ml/min respectively. A total of 24 ml of cells containing the culture broth were injected into the

carrier stream and mixed with an enzyme solution of 350 000 U/l of glucose oxidase (Fluka, Germany) of a volume of 36 ml. After passing a reaction coil 50 cm length, the oxygen uptake was measured using an oxygen electrode (ANASYSCON, Germany). To determine the oxygen consumed by cells only, no enzyme solution were injected. Calculating the difference of both dissolved oxygen peak heights, the glucose concentration can be determined. The time between sample taking and the measurement of the dissolved oxygen was $\Delta t = 45$ s.

For the automation of the FIA system, as well as glucose concentration determination, the software CAFCA (ANASYSCON, Germany) was applied. To reduce the measurement noise the continuous-discrete extended Kalman filter was used. This program was running on a separate PC and got the measurement results via a serial connection. A PI controller was applied to adjust the glucose concentration to the desired set point of 0.1 g/l [41].

The initial process conditions were [41]:

$t_0 = 6.68$ h, $X(t_0) = 1.25$ g/l, $S(t_0) = 0.8$ g/l, $S_{in} = 100$ g/l.

6.1.2 Optimization Criterion

From the practical perspective, modeling studies are performed to identify simple and easy-to-use models that are suitable to support the engineering tasks of process optimization and, especially, of control. The most appropriate model must satisfy the following conditions:

(i) the model structure should be able to represent the measured data in a proper manner;
(ii) the model structure should be as simple as possible, while remaining compatible with the first requirement.

On account of that, the cultivation process dynamic is described using a simple Monod-type model, the most common kinetics applied for modelling of the cultivation processes [38].

The optimization criterion is a certain factor, value of which defines the quality of an estimated set of parameters. To evaluate the mishmash between the experimental and the model predicted data, a modified Hausdorff Distance and the Least Square Regression are proposed.

6.1.2.1 Hausdorff Distance

When talking about distances, it usually means the shortest: for instance, if a point X is said to be at distance D of a polygon P, it is generally assumed that D is the distance from X to the nearest point of P. The same logic applies for polygons: if two polygons A and B are at some distance from each other, it commonly understood that the distance is the shortest one between any point of A and any point of B. That definition of distance between polygons can become quite unsatisfactory for some applications. However, it would be natural to expect that a small distance between two polygons means that no point of one polygon is far from the other polygon. Unfortunately, the shortest distance concept carries very low informative content.

In mathematics, the Hausdorff Distance, or the Hausdorff Metric (named after Felix Hausdorff), also called Pompeiu-Hausdorff Distance [33], measures how far two subsets of a metric space are from each other. It turns the set of non-empty compact subsets of a metric space into a metric space in its own right. Informally, two sets are close in the Hausdorff Distance if every point of either set is close to some point of the other set. In other words, the Hausdorff Distance is the longest distance you can be forced to travel by an adversary who chooses a point in one of the two sets, from where you then must travel to the other set. Thus, it is the farthest point of a set that you can be at, to the closest point of a different set. More formally, the Hausdorff Distance from set A to set B is a maxmin function defined as:

$$h(A, B) = \max_{a \in A} \left\{ \min_{b \in B} \{d(a, b)\} \right\}, \tag{6.6}$$

where a and b are points of sets A and B respectively, and $d(a, b)$ is any metric between these points. For simplicity, in this work, the $d(a, b)$ as the Euclidean distance between a and b is taken. If sets A and B are made of lines or polygons instead of single points, then $h(A, B)$ applies to all defining points of these lines or polygons, and not only to their vertices. The Hausdorff Distance gives an interesting measure of mutual proximity, by indicating the maximal distance between any point of one set to the other set. In this way it is better than the shortest distance, which applied only to one point of each set, irrespective of all other points of the sets.

6.1.2.2 Least Squares Regression

The objective of the modeling process consists of adjusting the parameters of a model function to best fit the data set. A simple data set consists of n points (data pairs) (x_i, y_i), $i = 1, 2, \ldots, n$, where x_i is an independent variable and y_i is a dependent variable value of which is found by observation. The model function has the form $f(x, \beta)$, where the m adjustable parameters are held in the vector β. The goal is to find the parameter values for the model which "best" fits the data. The least squares method finds its optimum when the sum S of squared residuals:

$$S = \sum_{i=1}^{n} r_i^2$$

is at a minimum. A residual is defined as the difference between the actual value of the dependent variable and the value predicted by the model. A data point may consist of more than one independent variable. For example, when fitting a plane to a set of height measurements, the plane is a function of two independent variables, x and z. In the most general case there may be one or more independent variables and one or more dependent variables at each data point.

$$r_i = y_i - f(x_i, \beta).$$

6.2 ACO for *E. coli* Cultivation Model

Here the proposed ACO approach is very close to real ant behavior. The parameters μ_{max}, k_S and $Y_{S/X}$ have to be estimated. First, the problem is represented by a graph. It is needed to find the optimal values of three parameters which are interrelated. Therefore, the problem is represented with three-partitive graph. The graph consists of three levels. Every level represents a search area of one of the parameters that will be optimized. Every area is thus discretized, to consists of 1000 points (nodes), which are uniformly distributed in the search interval of every parameter. The first level of the graph represents the parameter μ_{max}. The second level represents the parameter k_S. The third level represents the parameter $Y_{S/X}$. There are arcs between nodes from consecutive levels of the graph and there are no arcs between nodes from the same level. The pheromone is deposited on the arcs, to indicate how good is this parameter combination. Every level of the graph of the problem consists of 1000 points, thus the number of possible solutions is 10^9, therefore is unpractical to apply the exact methods. Starting to create a solution, the ants chose a node from the firs level in a random way. Next, for nodes from the second and the third level, they apply the probabilistic rule. The transition probability depends only on the pheromone level. The heuristic information is not used. Thus the transition probability is as follows:

$$p_{i,j} = \frac{\tau_{i,j}}{\sum\limits_{k \in Unused} \tau_{i,k}}, \tag{6.7}$$

The ants prefer the node with maximal probability, which is the node with maximal quantity of the pheromone on the arc (starting from the current node). If there is more than one candidate for next node, the ant chooses randomly between the candidates. The process is iterative. At the end of every iteration the pheromone on the arcs is updated. The quality of the solutions is represented by the value of the objective function. In this case the objective function is the mean distance between the simulated data and the experimental data, which are the concentration of the biomass and the concentration of the substrate. The aim of the process is to minimize it, therefore the new added pheromone by ant i is:

$$\Delta\tau = (1 - \rho)/J(i) \tag{6.8}$$

Table 6.1 Parameters of ACO algorithm

Parameter	Value
Number of ants	20
Initial pheromone	0.5
Evaporation	0.1

where $J(i)$ is the value of the objective function according the solution constructed by ant i. Thus the arcs corresponding to solutions with the lesser value of the objective function will receive more pheromone and will be more desirable in the next iteration.

The values of the parameters of the ACO algorithms are very important, because they manage the search process. Therefore, it is necessary to find appropriate parameter settings, where the number of ants is the main parameter. In the ACO a small number of ants between 10 and 20 can be used, without need to increase the number of iterations to achieve good solutions. The next parameter is the initial pheromone. Normally it has a small value. The last parameter is the evaporation rate, which shows the importance of the last found solution, as related to the previous ones. Parameters of the ACO were tuned based on several pre-tests according considered here optimization problem. After tuning procedures the main algorithm parameters are set to the optimal settings. The parameter settings for the ACO are shown in Table 6.1.

The modified Hausdorff Distance, which is conformable to the considered problem is applied. There are two sets of points, the simulated (model predicted) and the measured (experimental) data, which form two lines. The Euclidean distance $d(t)$ between points from the two lines, corresponding to the same time moment t, is calculated. After that, the Euclidean distance from the point of one of the lines in time t to the points from other line in the time interval $(t - d(t), t + d(t))$ is calculated, and the minimal of these distances is taken. This is the distance between the two lines in the time moment t. Thus, the number of calculations, compared with the traditional Hausdorff Distance, is decreased, due to the fact that the distance to the points out of the interval $(t - d(t), t + d(t))$ will be large. At the end all distances between the points and the lines are combined. Thereby, eventual larger distance in some time moment, due to the measurement noise, is eliminated.

Thus, the objective function is presented as a minimization of the modified Hausdorff Distance measure J_1 between experimental and model predicted values of the state variables, represented by the vector $\mathbf{y}$:

$$J_1 = \sum_{i=1}^{m} h\left(\mathbf{y}_{\text{exp}}(i), \mathbf{y}_{\text{mod}}(i)\right)^2 \rightarrow \min \tag{6.9}$$

where m is the number of state variables (biomass and glucose concentrations); $\mathbf{y}_{\text{exp}}$ is the known experimental data; while $\mathbf{y}_{\text{mod}}$ model predictions with a given set of the parameters.

In the case of the Least Square Regression the objective function is:

$$J_2 = \sum_{i=1}^{m} \left(\mathbf{y}_{\exp}(i) - \mathbf{y}_{\text{mod}}(i)\right)^2 \rightarrow \min \tag{6.10}$$

6.2.1 Numerical Results

In this subsection we report achieved results when the proposed ACO algorithm is applied with Least square regression and with Hausdorff distance as optimization criterion. ACO algorithm is compared with genetic (GA) algorithm All experiments have been conduced on a PC with Intel Core 2 2.8 GHz, 3.5 GB Memory, Linux operating system and using the Matlab 7.5 environment.

Because of the stochastic characteristics of the applied ACO and GA algorithms, a series of 30 runs for each algorithm was performed.

To study the algorithm performance, the worst the best and the average results of the 30 runs, for the objective function values of the two variants of ACO and GA algorithms are studied. For a realistic comparison, the number of iterations is fixed to be 100. The average, worst and best values of the objective functions are shown on Table 6.2. In the first line of the second row, the average, worst and best value of the objective function are shown when Least Square Regression is used with ACO. The second line in the second row depicts the calculated Hausdorff Distance between the same solutions achieved when the objective function is the Least Square Regression. The first line of the third row shows the average, best and worst values of the objective function when it is the Hausdorff Distance with ACO. The second line of the third row represents the Least Square Distance of the same solutions achieved when the objective function is the Hausdorff Distance. Comparing the two rows it can be observed that the average and the worst achieved results are much better using the Hausdorff Distance than the Least Square Regression. The best achieved solutions are similar. In the best achieved solutions it can be seen that the Hausdorff Distance between achieved solutions is smaller when the objective function is Hausdorff, but

Table 6.2 ACO and GA with Least Square Regression and Hausdorff Distance

	Method	Average	Worst	Best
ACO Least Square	–	4.8866	6.7700	3.3280
	Hausdorff	2.3875	4.1290	1.7218
ACO Hausdorff	–	1.8744	2.5322	1.6425
	Least Square	3.9706	4.4283	3.4276
GA Least Square	–	4.8341	5.4234	4.7314
	Hausdorff	1.7510	2.0224	1.7025
GA Hausdorff	–	2.0299	2.3326	1.7657
	Least Square	4.3549	4.8872	3.5464

Table 6.3 Best parameter values of the model

Parameter	Value
μ_{max}	0.5283
k_S	0.0174
$Y_{S/X}$	2.0300

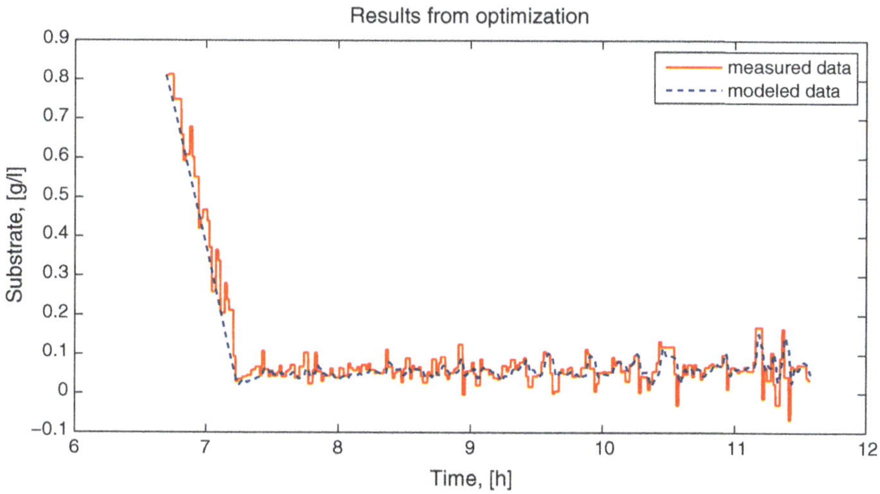

Fig. 6.1 Time profiles of the substrate: experimental data and models predicted data

the Least Square Distance is smaller when the objective function is the Least Squares Regression. During a number of runs of the algorithm the same phenomenon was observed—a small Hausdorff Distance between modeled and measured data and at the same time a big Least Square Distance between the data. When the Least Squares Regression is applied as the metric, the distance between the two lines can be very big, and in the same time it is seen that they are geometrically close to each other. It can happen especially in the steep parts of the lines. Applying the Hausdorff Metrics it can not happen, because it measures the geometrical similarity. Overall, the Hausdorff Distance is more time consuming than Least Square Distance, but much more realistic for the type of problems considered here. It can be concluded that ACO algorithm proposed in this paper performs better when the objective function is the Hausdorff Distance. We do the similar analysis when we apply GA. In this case there is very small difference between achieved results when we apply Hausdorff distance and Least Square Regression. We can conclude that GA is less sensitive according used measures.

In Table 6.3 the best parameter values (μ_{max}, k_S and $Y_{S/X}$), obtained using the ACO with the objective function based on the Hausdorff Distance, are presented.

The obtained model dynamics compared to the real experimental data is presented in Figs. 6.1 and 6.2.

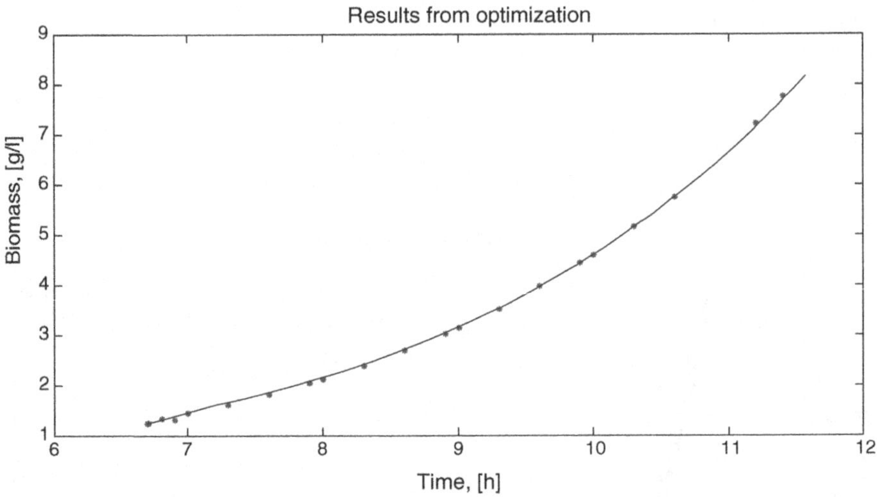

Fig. 6.2 Time profiles of the biomass: experimental data and models predicted data

In Fig. 6.1, the modelled substrate is represented by the dash line, while by solid line the measured substrate is depicted. In Fig. 6.2, line represents values of the modelled biomass, while stars represent values of the measured biomass.

The presented figures show a very good correlation between the experimental and model predicted data and confirm the obtained results.

6.3 Influence of the ACO Parameters on the Algorithm Performance

To find the best solution for reasonable computational time it is necessary to find appropriate ACO parameters values. The main ACO parameters that essential influence on algorithm performance are the number of ants and the number of iterations. Too many ants will quickly accumulate too many pheromone on the elements of suboptimal solutions and the algorithm will converge to the optimal solution very slowly. Using several ants only, will not produce enough pheromone accumulation and the algorithm convergence will be very slow too.

In order to investigate the influence of the number of ants, we propose a scheme where different number of ants and different number of iterations are combined to ensure constant algorithm running time. In our research we fixed the product of the number of ants and the number of iterations to be 2000. We run eight ACO algorithms with 5, 10, 20, 25, 40, 50, 100 and 200 ants for 400, 200, 100, 80, 50, 40, 20 and 10 iterations, respectively. The running time is the same for every combination of ants $\times$ iterations. The aim is to find the minimal number of ants we need to achieve good solution.

Table 6.4 ACO algorithms performance

ACO algorithm	Ants × iterations	5 × 400	10 × 200	20 × 100	25 × 80	40 × 50	50 × 40	100 × 20	200 × 10
Value of J	Average	**5.86**	6.25	6.67	6.30	5.91	6.27	**5.78**	6.60
	Best	3.49	3.43	3.70	3.52	3.43	3.45	3.43	3.68
	Worst	13.83	23.26	18.79	12.10	16.32	13.28	10.78	20.42

The initial pheromone and evaporation rate are another ACO parameters that should be defined according to the particular problem. The pheromone value lies between the ranges 0–1. Normally initial pheromone has a small value. The evaporation rate shows the importance of the last found solution as related to the previous ones. Usually the evaporation rate has a value between 0 and 1, too. These two ACO parameters were determined based on several pre-tests according to the considered here identification problem, as follows:

- initial pheromone = 0.5,
- evaporation rate = 0.1.

Due to the stochastic nature of the ACO we run each algorithm 30 times and calculate the average value of the objective function J. We apply ANOVA test to check whether the obtained results are statistically different. The results achieved by 5 ants and 400 iterations and by 100 ants and 20 iterations are statistically the same.

The resulting average, best and worst values of the objective function J for the considered eight ACO algorithms are reported in Table 6.4.

We observe that the best found results for all variants (number of ants) are very similar (statistically equivalent). Regarding the average results, the best performance is achieved of ACO with 5 ants × 400 iterations and ACO with 100 ants × 20 iterations.

The numerical results for objective function value J, over 30 runs of each ACO algorithm, are visualized on Fig. 6.3. We observe that the obtained values of J of ACO with 5 ants has less peaks compared to the ACO with 100 ants. The standard deviation of ACO with 5 ants × 400 iterations is less than of ACO with 100 ants × 20 iterations. Moreover more ants means more memory is needed. Therefore we can conclude that the best algorithm performance is when 5 ants and 400 iteration are used. Thus the minimal number of ants, necessary to achieve good solutions, is 5.

6.4 Hybryd ACO Algorithms

Most of the authors combine some metaheuristics with local search procedure or with some exact method. There are some applications of the hybrid algorithms between different metaheuristics. For example, the superiority of the hybrid algorithms between metaheuristics ACO and GA is shown in applications in different

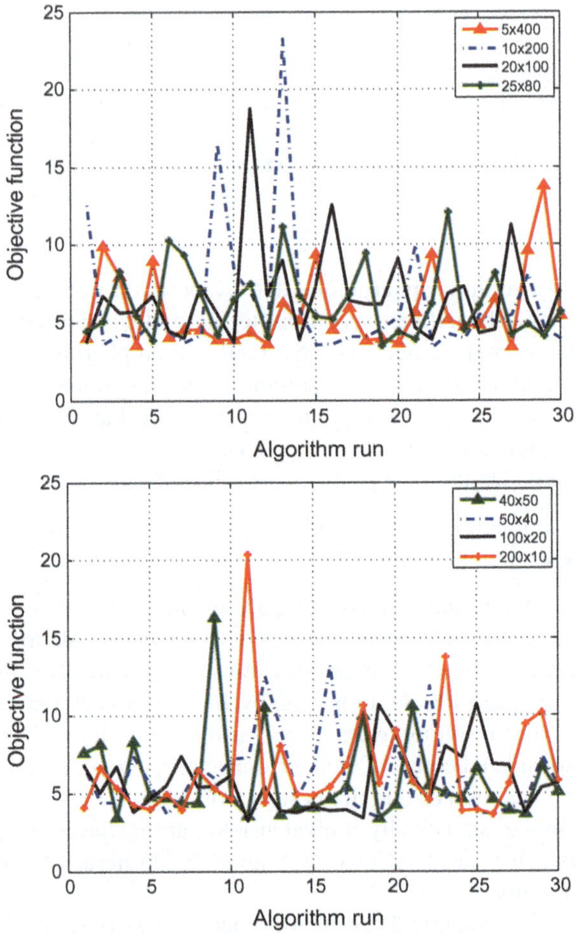

Fig. 6.3 Obtained objective function value J with the eight ACO algorithms

areas and problems [42–45]. In these papers first GA is applied and when it stagnates the ACO is used to go out of the stagnation and the algorithm continue with GA.

6.4.1 Hybrid ACO-GA Algorithms

In this subsection a hybrid metaheuristics ACO-GA is realized and applied for parameter identification of a cultivation process of the bacteria *E. coli* model. ACO is a constructive method which does not need initial solutions. GA is a population-based method and in traditional GA's initial population is randomly generated. In this random generation the initial solutions can be very far from the optimal solutions and may need a lot of iterations to draw close to it. Therefore, our idea is to generate initial

Table 6.5 Operators and parameters of GA

Operator	Type	Parameter	Value
Fitness function	Linear ranking	ggap	0.97
Selection function	Roulette wheel selection	xovr	0.75
Crossover function	Simple crossover	mutr	0.01
Mutation function	Binary mutation	maxgen	200
Reinsertion	Fitness-based	nind	180

Table 6.6 Parameters of ACO algorithm

Parameter	Value
Number of ants	5
Initial pheromone	0.5
Evaporation	0.1
Generations	200

solutions by ACO and then use them as initial population in GA. Thus, the GA will start with a population, which is closer to optimal solution. Sometimes, after a number of iterations, the GA goes to stagnation and the population stops being improved. So, our next idea is to return the GA solutions to the ACO algorithm and update the pheromone accordingly, then run the ACO with the updated pheromone, and thus to generate new population for the GA. In this way, we will perform hybridization in two directions: GA is hybridized with ACO and ACO is hybridized with GA. We can use any variant of ACO and GA, depending on the class of problems, which the algorithm is applied to. The best model parameters vector is obtained by the genetic evolution of ant colony.

Parameters of the GA and ACO were tuned based on several pre-tests according considered here optimization problem. After tuning procedures the main algorithm parameters are set to the optimal settings. The basic operators and parameters in GA are summarized in Table 6.5. The parameter setting for ACO is shown in Table 6.6.

We perform 30 independent runs of the three metaheuristics: ACO, GA and hybrid ACO-GA. Computer specification to run all identification procedures are Intel Core i5-2329 3.0 GHz, 8 GB Memory, Windows 7 (64bit) operating system and Matlab 7.5 environment. For compariseon of hybrid performance pure GA and ACO are run with parameters shown in Tables 6.5 and 6.14. Hybrid ACO-GA starts with 5 ants for 10 generation. To form population of chromosomes for further improvement from GA, pure ACO repeat 30 times. We take the best ACO solution from every one of the runs to form population. The obtained population with 30 chromosomes (ACO best solutions) is used from GA to obtain the best model parameters vector by the genetic evolution for 100 generations. The main numerical results are summarized in Table 6.7.

Table 6.7 Experimental results

Value	Algorithm	T, s	J
Best	GA	67,5172	4,4396
	ACO	67,3456	4,9190
	ACO-GA	35,5212	4,4903
Worst	GA	66,5968	4,6920
	ACO	66,6280	6,6774
	ACO-GA	35,3498	4,6865
Average	GA	67,1370	4,5341
	ACO	69,5379	5,5903
	ACO-GA	36,1313	4,5765

Table 6.7 shows that our ACO-GA algorithm achieves similar to pure ACO and pure GA solutions, but the running time is two times less. The pure GA algorithm starts from randomly generated initial solutions (population) which can be very fare from the optimal one. The ACO is a constructive method, which dos not need any initial solution. ACO can faster find solutions which are not very fare from the optimal one. In our hybrid ACO-GA algorithm we explore the advantages of the both ACO and GA. We run the ACO for several iterations only and thus we generate initial solutions for GA which are closer to the optimal. The GA starts from solutions which are not fare from the optimal and thus we increase the convergence of the GA. More over in ACO-GA the population is very small, only 5 ants for ACO and only 30 individuals for GA, which decreases the used memory. Thus our hybrid algorithm has two advantages—less running time and less memory.

6.4.2 Hybrid GA-ACO Algorithms

In this subsection we proposed to combine two metaheuristics, namely GA [16, 17] and ACO [20]. In our hybrid algorithm the solutions achieved by GA are like solutions achieved by ACO from some previous iteration and we update the initial pheromone according them. After that we continue with ACO algorithm. The pseudocode of the proposed GA-ACO algorithm is shown in Fig. 6.4.

To set the optimal settings of the GA and the ACO algorithms parameters, we performed several runs of the algorithms with varying parameters, according to the considered here optimization problem. The resulting optimal settings of the GA and the ACO parameters are summarized in Tables 6.8, 6.9 and 6.14.

The computer, used to run all identification procedures, was an Intel Core i5-2329 3.0 GHz, with 8 GB Memory, Windows 7 (64bit) operating system and Matlab 7.5 environment.

Fig. 6.4 Pseudocode for Hybrid GA-ACO

```
GA-ACO hybrid algorithm
i = 0
Initial population Pop(0)
Evaluate Pop(0)
while not end-condition do
        i = i + 1
        Select Pop(i) from Pop(i − 1)
        Recombine Pop(i)
        Mutate Pop(i)
        Evaluate Pop(i)
end while
Final GA solution for ACO
Initialize number of ants;
Initialize the ACO parameters;
while not end-condition do
        for k = 0 to number of ants
            ant k choses start node;
            while solution is not constructed do
                    ant k selects higher probability node;
            end while
        end for
        Update-pheromone-trails;
end while
Final solution
```

Table 6.8 Parameters of GA

Parameter	Value
gap	0.97
xovr	0.7
mutr	0.05
maxgen	200
Individuals	100
nvar	3
Inserted rate	100%

Table 6.9 Parameters of ACO algorithm

Parameter	Value
Number of ants	20
Initial pheromone	0.5
Evaporation	0.1
Generations	200

Table 6.10 Parameters of GA-ACO algorithm

Parameter	Value
gap	0.97
xovr	0.7
mutr	0.05
GA maxgen	40
Individuals	20
nvar	3
Inserted rate	100%
Number of ants	20
Initial pheromone	0.5
Evaporation	0.1
ACO generations	100

We performed 30 independent runs of the hybrid GA-ACO. The hybrid algorithm started with population of 20 chromosomes. We used 40 generations to find the initial solution. We took the achieved best GA solution to specify the ACO initial pheromones. Next, the ACO was used to obtain the best model parameters vector using 20 ants for 100 generations (see Table 6.10).

For comparison of performance of the hybrid algorithm we used the pure GA and the pure ACO. They were run (30 times) with (optimized) parameters shown in Tables 6.8 and 6.14.

The main numerical results, obtained when solving the parameter identification problem, are summarized in Table 6.11. In this table we show the best, worst and average values of the objective function achieved by the pure ACO, the pure GA and the hybrid GA-ACO algorithms after 30 run of every one of them, as well as their running times. The obtained average values of the model parameters (μ_{max}, k_S and $Y_{S/X}$) are summarized in Table 6.12.

As it can be seen, from Table 6.11, the hybrid GA-ACO achieves values of the objective function that are similar to these obtained by the pure GA and the pure ACO algorithms. In the same time, the running time of the proposed hybrid algorithm is about two times shorter. The pure ACO algorithm starts with an equal initial pheromone distribution for all problem elements. In the case of the hybrid GA-ACO we use the best solution found by the GA to specify the initial distribution of the pheromone (used by the ACO). Thus our ACO algorithm uses the GA "experience" and starts from a "better" pheromone distribution. This strategy helps the ants to find "good solutions" using less computational resources (e.g. like computer time and memory). As a matter of fact, our hybrid algorithm uses more than three times less memory than the pure ACO and the pure GA algorithms.

In Table 6.13 we compare results achieved in current subsection with results obtained in previous subsection. There, we had run the ACO algorithm for several iterations and used it to generate an initial populations for the GA algorithm. Thus

Table 6.11 Results from model parameters identification procedures

Value	Algorithm	Algorithm performance	
		T, [s]	J
Best	GA	67.5172	4.4396
	ACO	67.3456	4.9190
	GA-ACO	38.7812	4.3803
Worst	GA	66.5968	4.6920
	ACO	66.6280	6.6774
	GA-ACO	41.4495	4.6949
Average	GA	67.1370	4.5341
	ACO	69.5379	5.5903
	GA-ACO	39.4620	4.5706

Table 6.12 Parameters' estimations of the *E. coli* fed-batch fermentation process model

Value	Algorithm	Model parameters		
		μ_{max}	k_S	$1/Y_{S/X}$
Average	GA	0.4857	0.0115	2.0215
	ACO	0.5154	0.0151	2.0220
	GA-ACO	0.4946	0.0123	2.0204

Table 6.13 Results from model parameters identification procedures: ACO-GA

Value	ACO-GA performance	
	T, [s]	J
Best	35.5212	4.4903
Worst	41.4495	4.6865
Average	36.1313	4.5765

the GA started from a population that was closer to the good (optimal) solution than a randomly generated population. We observe that the ACO-GA and the GA-ACO algorithms achieve very similar results, and in a similar running time. We run the ANOVA test to measure the relative difference between the two algorithms. The two hybrid algorithms achieves statistically equivalent results, but the GA-ACO algorithm uses 30% less memory. Thus we can conclude that hybrid GA-ACO algorithm performs better than the ACO-GA hybrid algorithm.

6.5 InterCriteria Analysis of ACO Performance

For the model parameters identification we use experimental data for biomass and glucose concentration of an *E. coli* MC4110 fed-batch fermentation process.

To estimate the model parameters we applied consistently 11 differently tuned ACO algorithms. We use various ant numbers—from 5 to 100 ants, namely ACO_5, ACO_{10}, ACO_{20}, ..., ACO_{90}, ACO_{100}. The number of generations is fixed to 100. The main ACO parameters are summarized in Table 6.14.

Due to the stochastic nature of the applied algorithm we perform series of 30 runs for each differently tuned ACO algorithm. Thus, we obtain the average, best and worst estimate of the parameters, as well as of the algorithm execution time and value of objective function. The detailed description of identification procedure is given in [46].

To perform ICrA three IMs are constructed—the IM A_1 (Eq. 6.11) with the obtained average results, the IM A_2 (Eq. 6.12) with the best obtained results and IM A_3 (Eq. 6.13) with the worst obtained results.

$$A_1 = \begin{array}{r|ccccc} & C_1 & C_2 & C_3 & C_4 & C_5 \\ \hline ACO_5 & 0.5700 & 0.0285 & 2.0260 & 6.2523 & 16.9105 \\ ACO_{10} & 0.5436 & 0.0237 & 2.0180 & 6.0527 & 29.4062 \\ ACO_{20} & 0.4893 & 0.0148 & 2.0261 & 5.4330 & 56.0980 \\ ACO_{30} & 0.4968 & 0.0116 & 2.0320 & 5.2849 & 90.6210 \\ ACO_{40} & 0.5112 & 0.0180 & 2.0262 & 5.2853 & 109.6219 \\ ACO_{50} & 0.5148 & 0.0156 & 2.0263 & 5.2206 & 131.3684 \\ ACO_{60} & 0.4962 & 0.0127 & 2.0220 & 5.2184 & 151.6018 \\ ACO_{70} & 0.5049 & 0.0171 & 2.0180 & 5.1350 & 173.7539 \\ ACO_{80} & 0.4719 & 0.0105 & 2.0280 & 5.1324 & 197.6533 \\ ACO_{90} & 0.5082 & 0.0152 & 2.0221 & 5.1415 & 237.5271 \\ ACO_{100} & 0.4737 & 0.0108 & 2.0240 & 5.0885 & 260.1005 \end{array} \tag{6.11}$$

Table 6.14 Parameters of ACO algorithm

Parameter	Value
Number of ants (nind)	5–100
Initial pheromone	0.5
Evaporation	0.1
Maximum generations (maxgen)	100

$$A_2 = \begin{array}{c|ccccc} & C_1 & C_2 & C_3 & C_4 & C_5 \\ \hline ACO_5 & 0.4911 & 0.0139 & 2.0200 & 5.0652 & 16.7857 \\ ACO_{10} & 0.4986 & 0.0130 & 2.0220 & 4.8083 & 29.7494 \\ ACO_{20} & 0.4806 & 0.0116 & 2.0260 & 4.9293 & 55.5208 \\ ACO_{30} & 0.4956 & 0.0146 & 2.0221 & 4.7408 & 90.7302 \\ ACO_{40} & 0.4794 & 0.0101 & 2.0262 & 4.8004 & 109.4347 \\ ACO_{50} & 0.4959 & 0.0145 & 2.0201 & 4.6598 & 131.4308 \\ ACO_{60} & 0.4854 & 0.0110 & 2.0263 & 4.8983 & 152.1166 \\ ACO_{70} & 0.4977 & 0.0114 & 2.0222 & 4.7739 & 173.7383 \\ ACO_{80} & 0.4827 & 0.0109 & 2.0240 & 4.8078 & 196.7641 \\ ACO_{90} & 0.4800 & 0.0100 & 2.0241 & 4.7856 & 234.3915 \\ ACO_{100} & 0.4884 & 0.0126 & 2.0261 & 4.8382 & 260.1473 \end{array} \qquad (6.12)$$

$$A_3 = \begin{array}{c|ccccc} & C_1 & C_2 & C_3 & C_4 & C_5 \\ \hline ACO_5 & 0.5532 & 0.0183 & 2.0120 & 7.9011 & 16.7857 \\ ACO_{10} & 0.5115 & 0.0146 & 2.0040 & 8.0956 & 29.6246 \\ ACO_{20} & 0.5733 & 0.0278 & 2.0140 & 6.5924 & 55.5052 \\ ACO_{30} & 0.4641 & 0.0104 & 2.0240 & 6.2202 & 90.7146 \\ ACO_{40} & 0.5292 & 0.0148 & 2.0221 & 6.0784 & 112.8667 \\ ACO_{50} & 0.5043 & 0.0169 & 2.0160 & 5.7156 & 131.9924 \\ ACO_{60} & 0.5376 & 0.0198 & 2.0161 & 5.7759 & 151.1026 \\ ACO_{70} & 0.5232 & 0.0161 & 2.0200 & 5.6652 & 175.1891 \\ ACO_{80} & 0.4746 & 0.0125 & 2.0220 & 5.7891 & 199.2601 \\ ACO_{90} & 0.5187 & 0.0211 & 2.0180 & 5.6120 & 236.0919 \\ ACO_{100} & 0.4881 & 0.0122 & 2.0340 & 5.4866 & 258.4781 \end{array} \qquad (6.13)$$

In addition to the presented in [47] results here the average, worst and best estimates for the three model parameters in all 11 cases are given too. Thus, five criteria are considered—C_1 is the parameter μ_{max}, C_2 is the parameter k_S, C_3 is the parameter $Y_{S/X}$, C_4 is the objective function value J and C_5 is the resulting execution time T.

Based on the presented **Algorithm 1** the ICrA is implemented in the Matlab 7.5 environment. We obtain IMs that determine the degrees of "agreement" ($\mu_{C,C'}$) and "disagreement" ($\nu_{C,C'}$) between criteria for the average, worst and best ACO results.

6.5.1 Average Results

Resulting degrees of "agreement" ($\mu_{C,C'}$) (IM_1) and degrees of "disagreement" ($\nu_{C,C'}$) (IM_2) are as follows:

$$IM_1 = \begin{array}{c|ccccc} & C_1 & C_2 & C_3 & C_4 & C_5 \\ \hline C_1 & \mathbf{1} & 0.89 & 0.42 & 0.76 & 0.29 \\ C_2 & 0.89 & \mathbf{1} & 0.35 & 0.76 & 0.25 \\ C_3 & 0.42 & 0.35 & \mathbf{1} & 0.47 & 0.49 \\ C_4 & 0.76 & 0.76 & 0.47 & \mathbf{1} & 0.05 \\ C_5 & 0.29 & 0.25 & 0.49 & 0.05 & \mathbf{1} \end{array}, \quad IM_2 = \begin{array}{c|ccccc} & C_1 & C_2 & C_3 & C_4 & C_5 \\ \hline C_1 & \mathbf{0} & 0.11 & 0.56 & 0.24 & 0.71 \\ C_2 & 0.11 & \mathbf{0} & 0.64 & 0.24 & 0.75 \\ C_3 & 0.56 & 0.64 & \mathbf{0} & 0.51 & 0.49 \\ C_4 & 0.24 & 0.24 & 0.51 & \mathbf{0} & 0.95 \\ C_5 & 0.71 & 0.75 & 0.49 & 0.95 & \mathbf{0} \end{array} \tag{6.14}$$

ICrA analysis of the average results shows some degrees of "uncertainty" ($\pi_{C,C'}$, Eq. (6.4) as follows:

$$IM_3 = \begin{array}{c|ccccc} & C_1 & C_2 & C_3 & C_4 & C_5 \\ \hline C_1 & \mathbf{0} & 0 & 0.02 & 0 & 0 \\ C_2 & 0 & \mathbf{0} & 0.02 & 0 & 0 \\ C_3 & 0.02 & 0.02 & \mathbf{0} & 0.02 & 0.02 \\ C_4 & 0 & 0 & 0.02 & \mathbf{0} & 0 \\ C_5 & 0 & 0 & 0.02 & 0 & \mathbf{0} \end{array} \tag{6.15}$$

6.5.2 Worst Results

Resulting degrees of "agreement" ($\mu_{C,C'}$) (IM_4) and degrees of "disagreement" ($\nu_{C,C'}$) (IM_5) are as follows:

$$IM_4 = \begin{array}{c|ccccc} & C_1 & C_2 & C_3 & C_4 & C_5 \\ \hline C_1 & \mathbf{1} & 0.82 & 0.31 & 0.60 & 0.35 \\ C_2 & 0.82 & \mathbf{1} & 0.27 & 0.49 & 0.45 \\ C_3 & 0.31 & 0.27 & \mathbf{1} & 0.31 & 0.75 \\ C_4 & 0.60 & 0.49 & 0.31 & \mathbf{1} & 0.09 \\ C_5 & 0.35 & 0.45 & 0.75 & 0.09 & \mathbf{1} \end{array}, \quad IM_5 = \begin{array}{c|ccccc} & C_1 & C_2 & C_3 & C_4 & C_5 \\ \hline C_1 & \mathbf{0} & 0.18 & 0.69 & 0.40 & 0.65 \\ C_2 & 0.18 & \mathbf{0} & 0.73 & 0.51 & 0.55 \\ C_3 & 0.69 & 0.73 & \mathbf{0} & 0.69 & 0.25 \\ C_4 & 0.40 & 0.51 & 0.69 & \mathbf{0} & 0.91 \\ C_5 & 0.65 & 0.55 & 0.25 & 0.91 & \mathbf{0} \end{array} \tag{6.16}$$

6.5.3 Best Results

Resulting degrees of "agreement" ($\mu_{C,C'}$) (IM_6) and degrees of "disagreement" ($\nu_{C,C'}$) (IM_7) are as follows:

$$IM_6 = \begin{array}{c|ccccc} & C_1 & C_2 & C_3 & C_4 & C_5 \\ \hline C_1 & \mathbf{1} & 0.78 & 0.27 & 0.44 & 0.42 \\ C_2 & 0.78 & \mathbf{1} & 0.27 & 0.51 & 0.31 \\ C_3 & 0.27 & 0.27 & \mathbf{1} & 0.62 & 0.71 \\ C_4 & 0.44 & 0.51 & 0.62 & \mathbf{1} & 0.40 \\ C_5 & 0.42 & 0.31 & 0.71 & 0.40 & \mathbf{1} \end{array}, \quad IM_7 = \begin{array}{c|ccccc} & C_1 & C_2 & C_3 & C_4 & C_5 \\ \hline C_1 & \mathbf{0} & 0.22 & 0.73 & 0.56 & 0.58 \\ C_2 & 0.22 & \mathbf{0} & 0.73 & 0.49 & 0.69 \\ C_3 & 0.73 & 0.73 & \mathbf{0} & 0.38 & 0.29 \\ C_4 & 0.56 & 0.49 & 0.38 & \mathbf{0} & 0.60 \\ C_5 & 0.58 & 0.69 & 0.29 & 0.60 & \mathbf{0} \end{array} \tag{6.17}$$

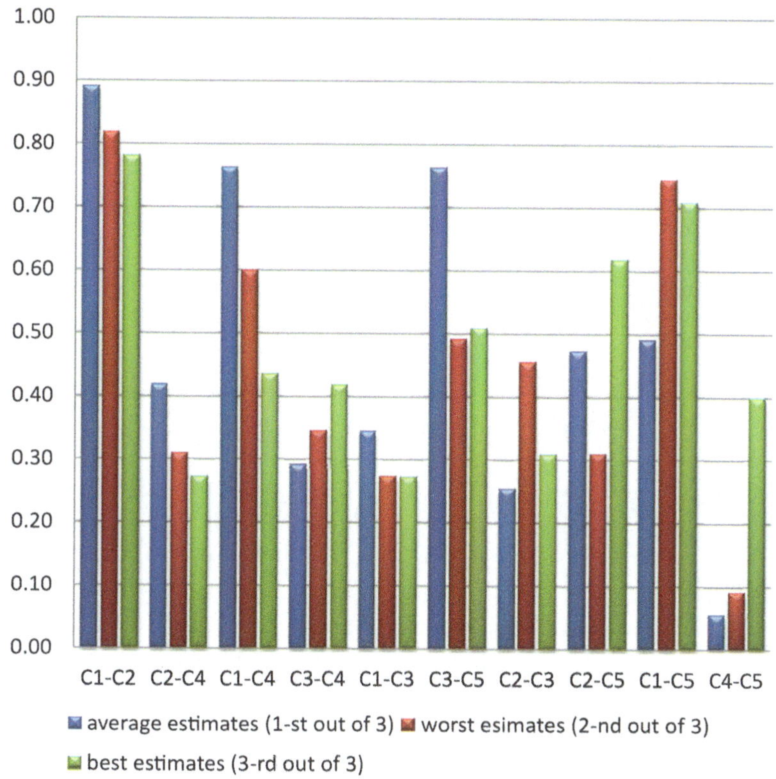

Fig. 6.5 Degrees of "agreement" ($\mu_{C,C'}$ values) for all cases

ICrA analysis of the best and the worst results do not shows degrees of "uncertainty", i.e. $\pi_{C,C'} = 0$ for all criteria pairs.

The obtained degrees of "agreement" ($\mu_{C,C'}$ values) between considered criteria for average, best and worst results are presented in Fig. 6.5.

Let us consider the following scheme for defining the consonance and dissonance between each pair of criteria (see Table 6.15), where SNC is strong negative consonance, NC is negative consonance, WNC is weak negative consonance, WD is weak dissonance, D is dissonance, SD is strong dissonance, WPC is weak positive consonance, PC is positive consonance and SPC is strong positive consonance [48].

In Table 6.15 the results obtained in this research are compared to the results presented in [49]. In [49] for parameters identification of an *E. coli* MC4110 FP model consistently 14 differently tuned GA are applied. Various population sizes—from 5 to 200 chromosomes in the population are used. The number of generations is fixed to 200. The same five criteria are defined: C_1 is the parameter μ_{max}, C_2 is the parameter k_S, C_3 is the parameter $Y_{S/X}$, C_4 is the objective function value J and C_5 is the resulting execution time T.

Table 6.15 Criteria relations sorted by $\mu_{C,C'}$ values

$\mu_{C,C'}$	Meaning	Average results		Best results		Worst results	
		ACO	GA [49]	ACO	GA [49]	ACO	GA [49]
[0–0.05]	SNC	C_5-C_4					
(0.05-0.15]	NC		C_5-C_4			C_5-C_4	C_5-C_3 C_5-C_4
(0.15–0.25]	WNC	C_2-C_3			C_5-C_4		C_2-C_3 C_4-C_3
(0.25–0.33]	WD	C_5-C_3	C_2-C_3 C_5-C_3	C_2-C_3 C_4-C_3 C_2-C_4	C_5-C_2	C_3-C_4 C_3-C_1 C_2-C_3	C_5-C_1
(0.33–0.43]	D	C_2-C_4 C_4-C_3	C_5-C_2 C_2-C_4 C_4-C_3	C_5-C_3 C_5-C_4	C_5-C_3 C_4-C_3	C_5-C_1	C_2-C_4
(0.43–0.57]	SD	C_5-C_2 C_5-C_1	C_5-C_1	C_1-C_3 C_1-C_4	C_1-C_3 C_2-C_3 C_2-C_4	C_2-C_4 C_2-C_5	
(0.57–0.67]	D			C_5-C_1	C_1-C_4	C_1-C_4	C_5-C_2
(0.67–0.75]	WD		C_1-C_4	C_5-C_2	C_1-C_2 C_5-C_1	C_5-C_3	
(0.75–0.85]	WPC	C_1-C_4 C_1-C_3	C_1-C_3	C_1-C_2		C_1-C_2	C_1-C_2 C_1-C_3
(0.85–0.95]	PC	C_1-C_2	C_1-C_2				C_1-C_4
(0.95–1]	SPC						

Analysis of the both average results (here presented and the average results presented in [49]) shows the following criteria pair relations:

- For the pair C_5-C_4 (i.e., $T \leftrightarrow J$) a negative consonance is identified for average and worst results. Such dependence is logical—for a large number of algorithm iterations (i.e., greater execution time T) it is more likely to find a more accurate solution, i.e. a small value of J.
- For most results we established strong correlation between criteria C_1-C_2 (i.e., $\mu_{max} \leftrightarrow k_S$) and C_1-C_3 (i.e., $\mu_{max} \leftrightarrow Y_{X/S}$). There are two exceptions—GA best results and ACO worst results, where the criteria pair C_1-C_3 is respectively in SD and WD. Considering the physical meaning of the model parameters [38] it is clear that there is dependence between these criteria. A strong correlation is expected between criteria C_1-C_2 [38]. Considering relation C_1-C_3 we found some exceptions—GA best and ACO best and worst results.

Due to stochastic nature of considered here meta-heuristic techniques (GA and ACO) we observed some different criteria dependences based on the ICrA of the worst and best results:

- Based on the worst ACO and GA results we found weaker relation between criteria pairs C_1-C_5, C_2-C_4 and C_2-C_5. There are some small discrepancies for the pairs C_3-C_4 and C_2-C_3, and some larger discrepancies—for the pairs C_1-C_3, C_1-C_4, and C_5-C_3. For the pairs C_1-C_4, C_1-C_2 and C_1-C_3 we observed higher value of $\mu_{C,C'}$, i.e. PC or WPC.
- Compared to the ICrA of average results there are some discrepancies too. For example, average results show that the pair C_4-C_3 is in dissonance, while worst results—in WD (ACO results) or WNC (GA results).
- In the ICrA of the best results we identify the same results—some discrepancies are observed. There are some small discrepancies for the pairs C_3-C_4, C_1-C_4, C_2-C_1, C_2-C_4, C_5-C_4 and C_5-C_1, and some larger discrepancies—for the pairs C_2-C_3 and C_5-C_2. Here we observed a higher value of $\mu_{C,C'}$ only for the pair C_1-C_2, i.e. WPC (ACO results).

Taking into account the nature of the GA and ACO we consider that the ICrA of the average results has the highest significance.

6.6 Conclusion

In this chapter we propose an ACO algorithm for parameter identification of the *E. coli* fed-batch fermentation process. Two variants of objective function is proposed: least square regression and Hausdorff distance. The Hausdorff distance give more precise solution, but is more time consuming according least square regression. Influence of the ACO parameters on algorithm performance has been investigated. We look for a minimal number of ants that are needed to find good solution. We propose a two variants of hybrid algorithms combining GA and ACO in a different way. In one of the variants firs we apply ACO for several iterations only and achieved by ACO solutions we use as initial population for GA. Thus the GA starts from closer to the optimal solutions instead to start from random solutions. It leads to used of less population and less iterations, which means less computational resources. In another variant of hybrid algorithm we start the GA for several generations with a small population. Next, we use the best solution found by the GA, to instantiate the initial pheromone distribution for the ACO algorithm. We observe that our hybrid GA-ACO algorithm achieves results similar to the pure GA and the pure ACO algorithms, but it is using less computational resources (time and memory). The used time is two times smaller while the used memory is three times smaller. With this algorithm we understand how important is the pheromone distribution for good performance of the ACO algorithm. We compare our hybrid GA-ACO approach, with a hybrid ACO-GA algorithm. Both hybrid algorithms achieve statistically similar results for a similar running time, but GA-ACO algorithm uses about 30% less memory, which is important when one is to solve large problems.

Based on the apparatus of the Index Matrices and the Intuitionistic Fuzzy Sets, InterCriteria Analysis of a model parameters identification using Ant Colony Optimization is performed. A non-linear model of an *E. coli* fed-batch fermentation process is considered. Series of model identification procedures using Ant Colony Optimization are done. The InterCriteria Analysis is applied to explore the existing relations and dependencies of defined model parameters and Ant Colony Optimization outcomes—execution time and objective function value. Three case studies are examined considering average, worst and best results for the obtained model parameters, execution time and objective function value. The obtained results are compared to the results from a model parameters identification using Genetic Algorithms. Applying the InterCriteria Analysis and analyzing the results we establish relations and dependencies between the defined criteria. Based on the used scale for defining the consonance and dissonance between each pair of criteria, we discuss which criteria are in consonance and dissonance, as well as the degree of their dependence.

References

1. Fidanova, S., Roeva, O., Ganzha, M.: ACO for parameter settings of E. coli fed-batch cultivation model. In: Proceedings of FedCSIS 2012, pp. 407–414 (2012). IEEE Xplorer
2. Fidanova, S., Paprzycki, M., Roeva, O.: Hybrid GA-ACO algorithm for a model parameter identification problem. In: Proceedings of FedCSIS 2014 Conference, pp. 413–420 (2014). IEEE Xplorer
3. Fidanova, S., Roeva, O.: Influence of ant colony optimization parameters on the algorithm performance. Large Scale Sciantific Comput. Lect. Notes Comput. Sci. **10665**, 358–365 (2018). Springer
4. Roeva, O., Fidanova, S.: Metaheuristic techniques for optimization of an *E. coli* cultivation model. J. Biotechnol. & Biotechnol. Equipment **27**(3), 3870–3876 (2013)
5. Roeva, O., Fidanova, S., Atanassova, V.: Hybrid ACO-GA for parameter identification of an E. coli cultivation process model. Large Scale Sci. Comput. Lect. Notes Comput. Sci. **8353**, 288–295 (2014). Springer
6. Roeva, O., Fidanova, S.: Parameter identification of an *E.coli* cultivation process model using hybrid metahaeuristics. J. Metaheuristics **3**(2), 133–148 (2014)
7. Viesturs, U., Karklina D., Ciprovica, I.: Bioprocess and Bioengineering. Jeglava (2004)
8. Schuegerl, K., Bellgardt, K.-H.: Bioreaction Engineering: Modeling and Control. Springer-Verlag, Berlin, Heidelberg (2000)
9. Pardalos, P.M., Resende, M.G.C.: Handbook of Applied Optimization. Oxford University Press (2002)
10. Covert, M.W., Xiao, N., Chen, T.J., Karr, J.R.: Integrating metabolic, transcriptional regulatory, and signal transduction models in Escherichia coli. J. of Bioinform. **24**(18), 2044–2050 (2008)
11. Jiang, L., Ouyang, Q., Tu, Y.: Quantitative modeling of *Escherichia coli* chemotactic motion in environments varying in space and time. PLoS Comput. Biol. **6**(4) (2010). https://doi.org/10.1371/journal.pcbi.1000735
12. Karelina, T.A., Ma, H., Goryanin, I., Demin, O.V.: EI of the phosphotransferase system of Escherichia coli: Mathematical modeling approach to analysis of its kinetic properties. J. Biophys. **579402** (2011). https://doi.org/10.1155/2011/579402
13. Opalka, N., Brown, J., Lane, W.J., Twist, K.-A.F., Landick, R., Asturias, F.J., Darst, S.A.: Complete structural model of *Escherichia coli* RNA polymerase from a hybrid approach. PLoS Biol. **8**(9) (2010). https://doi.org/10.1371/journal.pbio.1000483
14. Petersen, C.M., Rifai, H.S., Villarreal, G.C., Stein, R.: Modeling Escherichia coli and its sources in an urban bayou with hydrologic simulation program - FORTRAN. J. Environ. Eng. **137**(6), 487–503 (2011)

15. Skandamis, P.N., Nychas, G.E.: Development and evaluation of a model predicting the survival of Escherichia coli O157:H7 NCTC 12900 in homemade eggplant salad at various temperatures, pHs, and oregano essential oil concentrations. Appl. Environ. Microbiol. **66**(4), 1646–1653 (2000)
16. Goldberg, D.E.: Genetic Algorithms in Search, Optimization and Machine Learning. Addison Wesley Longman, London (2006)
17. Holland, J.H.: Adaptation in Natural and Artificial Systems, 2nd edn. MIT Press, Cambridge (1992)
18. Michalewicz, Z.: Genetic Algorithms + Data Structures = Evolution Programs, 2 exended edn. Springer-Verlag, Berlin, Heidelberg (1994)
19. Dorigo, M., Di Caro, G.: The ant colony optimization meta-heuristic. In: Corne, D, Dorigo, M., Glover, F. (eds.) New Idea in Optimization. McGrow-Hill, pp. 11–32 (1999)
20. Dorigo, M., Stutzle, S.: Ant Colony Optimization. MIT Press (2004)
21. Fidanova, S.: ACO Algorithm with additional reinforcement. Int. Conf. Ant Colonies Artif. Ants Lect. Notes Comput. Sci. **2463**, 292–293 (2002). Springer
22. Fidanova, S., Alba, E., Molina, G.: Hybrid ACO algorithm for the GPS surveying problem. Large Scale Sci. Comput. Lect. Notes Comput. Sci. **5910**, 318–325 (2010). Springer
23. Umarani, R., Selvi, V.: Particle swarm optimization: Evolution, overview and applications. Int. J. Eng. Sci. Technol. **2**(7), 2802–2806 (2010)
24. Yusof, M.K., Stapa, M.A.: Achieving of Tabu search algorithm for scheduling technique in grid computing using GridSim simulation tool: Multiple jobs on limited resource. Int. J. Grid Distrib. Comput. **3**(4), 19–31 (2010)
25. Kirkpatrick, S., Gelatt, C.D., Vecchi, M.P.: Optimization by simulated annealing. Sci. New Ser. **220**(4598), 671–680 (1983)
26. Bonabeau, E., Dorigo, M., Theraulaz, G.: Swarm Intelligence: From Natural to Artificial Systems. Oxford University Press, New York (1999)
27. Syam, W.P., Al-Harkan, I.M.: Comparison of three meta heuristics to optimize hybrid flow shop scheduling problem with parallel machines. World Acad. Sci. Eng. Technol. **62**, 271–278 (2010)
28. Tahouni, N., Smith, R., Panjeshahi, M.H.: Comparison of stochastic methods with respect to performance and reliability of low-temperature gas separation processes. Can. J. Chem. Eng. **88**(2), 256–267 (2010)
29. Brownlee, J.: Clever algorithms. Nature-Inspired Programming Recipes, LuLu, p. 436 (2011). ISBN 978-1-4467-8506-5
30. Fidanova, S., Lirkov, I.: 3D protein structure prediction. J. Analele Universitatii de Vest Timisoara, Seria Matematica-Informatica **XLVII**(2), 33–46 (2009). ISSN 1224-970X
31. Fidanova, S.: An improvement of the grid-based hydrophobic-hydrophilic model. Int. J. Bioautoma. **14**(2), 147–156 (2010). ISSN 1312-451X
32. Shmygelska, A., Hoos, H.H.: An ant colony optimization algorithm for the 2D and 3D hydrophobic polar protein folding problem. BMC Bioinf. **6**(30) (2005). https://doi.org/10.1186/1471-2105-6-30
33. Rote, G.: Computing the minimum Hausdorff distance between two point sets on a line under translation. Inf. Process. Lett. **38**, 123–127 (1991)
34. Yedjour, H., Meftah, B., Yedjour, D., Benyettou, A.: Combining spiking neural network with Hausdorff Distance matching for object tracking. Asian J. App. Sci. **4**(2011), 63–71 (2011)
35. Sugiyama, M., Hirowatari, E., Tsuiki, H., Yamamoto, A.: Learning figures with the hausdorff metric by fractals. In: Proceedings of the 21st International Conference on Algorithmic Learning Theory. Springer-Verlag Berlin, Heidelberg, pp. 315–329 (2010)
36. Nutanong, S., Jacox, E.H., Samet, H.: An incremental Hausdorff Distance calculation algorithm. Proc. VLDB Endowment **4**(8), 506–517 (2011)
37. Chen, S., Lovell B.C.: Feature space Hausdorff distance for face recognition. In: Proceedings of 20th International Conference on Pattern Recognition (ICPR), Istanbul, Turkey, pp. 1465–1468 (2010)

38. Bastin, G., Dochain, D.: On-line Estimation and Adaptive Control of Bioreactors. Elsevier, Amsterdam (1991)
39. Roeva, O.: Parameter estimation of a monod-type model based on genetic algorithms and sensitivity analysis. Lect. Notes Comput. Sci. Springer-Verlag, Berlin Heidelberg **4818**, 601–608 (2008)
40. Roeva, O., Pencheva, T., Hitzmann, B., Tzonkov, S.: A genetic algorithms based approach for identification of *Escherichia coli* fed-batch fermentation. Int. J. Bioautomation. **1**, 30–41 (2004)
41. Arndt, M., Hitzmann, B.: Feed forward/feedback control of glucose concentration during cultivation of *Escherichia coli*. 8th IFAC Int. Conf. on Comp. Appl. in Biotechn. Canada, pp. 425–429 (2001)
42. Basiri, M.E., Nemati, S.: A novel hybrid ACO-GA algorithm for text feature selection. IEEE Congress on Evolutionary Computation, CEC '09, pp. 2561–2568 (2009)
43. Guangdong, H., Qun, W.: A hybrid ACO-GA on sports competition scheduling. In: Ostfeld, A. (ed.) Ant Colony Optimization - Methods and Applications, pp. 89–100. InTech (2011)
44. Li, N., Wang, S., Li, Y.: A hybrid approach of GA and ACO for VRP. J. Comput. Inf. Sys. **7**(13), 4939–4946 (2011)
45. Nemati, S., Basiri, M.E., Ghasem-Aghaee, N., Aghdam, M.H.: A novel ACO-GA hybrid algorithm for feature selection in protein function prediction. J. Expert Sys. Appl. An Int. J. Arch. **36**(10), 12086–12094 (2009)
46. Roeva, O., Fidanova, S., Paprzycki, M.: InterCriteria analysis of ACO and GA hybrid algorithms. Stud. Computat. Intell. **610**, 107–126 (2016)
47. Roeva, O., Fidanova, S., Paprzycki, M.: Influence of the population size on the geneticalgorithm performance in case of cultivation process modelling. In: Proceedings of the Federated Conference on Computer Science and Information Systems (FedCSIS), WCO 2013, Poland, pp. 371–376 (2013)
48. Atanassov, K., Atanassova, V., Gluhchev, G.: InterCriteria analysis: Ideas and problems. Notes on Intuitionistic Fuzzy Sets **21**(1), 81–88 (2015)
49. Roeva, O., Fidanova, S., Vassilev, P., Gepner, P.: InterCriteria analysis of a model parameters identification using genetic algorithm. Ann. Comput. Sci. Inf. Sys. **5**, 501–506 (2015)

Chapter 7
ACO for Wireless Sensor Network Positioning

The text of this chapter is based on [1–8]. Spatially distributed sensors, which communicate wirelessly form a wireless sensor network (WSN). This network monitors physical or environmental conditions. A central gateway, called high energy communication node, collects data from all sensors and sends them to the central computer where they are processed. We need to minimize the number of sensors and energy consumption of the network, when the terrain is fully covered. The problem is multi-objective. We propose ACO algorithm for solving the problem like multi-objective and two variants like mono-objective.

7.1 Introduction

The development of new technologies during the last decades gives a possibility for wireless data transmission. Thus a new types of networks, called wireless networks, was created.

Wireless Sensor Networks allow the monitoring of large areas without the intervention of a human operator. Their working is based on the exchange of local information between nodes in order to achieve a global goal. Cooperation between the sensor nodes is an important feature when solving complex tasks.

The WSN can be used in areas where traditional networks fail or are inadequate. They find applications in a variety of areas such as climate monitoring, military use, industry and sensing information from inhospitable locations. Unlike other networks, sensor networks depend on deployment of sensors over a physical location to fulfill a desired task. Sometimes deployments imply the use of hundreds or thousands of sensor nodes in small areas to ensure effective coverage range in a geographical field.

For a lot of applications wireless sensors offer a lower-cost method for collecting system health data to reduce energy usage and better manage resources. Wireless sensors are used to effectively monitor highways, bridges, and tunnels. Other applications are continually monitor office buildings, hospitals, airports, factories, power

S. Fidanova, *Ant Colony Optimization and Applications*, Studies in Computational Intelligence 947, https://doi.org/10.1007/978-3-030-67380-2_7

plants, or production facilities. The sensors can sens temperature, voltage, or chemical substances. A WSN allows automatically monitoring of almost any phenomenon. The WSN gives a lot of possibilities and offers a great amount of new problems to be solved. WSN have been used in military activities such es reconnaissance, surveillance [9], environmental activities such as forest fire prevention, geophysical activities such as volcano eruptions study [10], health data monitoring [11] or civil engineering [12].

A WSN node contains several components including the radio, battery, microcontroller, analog circuit, and sensor interface. In battery-powered systems, higher data rates and more frequent radio use consume more power. There are several open issues for sensor networks such as signal processing [13], deployment [14], operating cost, localization and location estimation.

The wireless sensors, have two fundamental functions: sensing and communicating. The sensing can be of different types (seismic, acoustic, chemical, optical, etc.). However, the sensors which are fare from the high energy communication node (HECN) can not communicate with him directly. Therefore the sensors transmit their data to this node, either directly or via hops, using nearby sensors as communication relays.

When deploying a WSN, the positioning of the sensor nodes becomes one of the major concerns. The coverage obtained with the network and the economic cost of the network depend directly of it. Since many WSN can have large numbers of nodes, the task of selecting the geographical positions of the nodes for an optimally designed network can be very complex. Therefore, metaheuristics seem an interesting option to solve this problem.

In this chapter we propose an algorithm which solves the WSN layout problem using ACO. We focus on both minimizing the energy depletion of the nodes in the network and minimizing the number of the nodes, while the full coverage of the network and connectivity are considered as constraints. The problem is multi-objective. We convert it to mono-objective in a two way: by multiplying the objective functions and by weighted sum of the objective functions. We compare achieved solutions with a multi-objective variant of the problem. We learn the algorithm performance and influence of the number of ants on achieved solutions.

Jourdan [15] solved an instance of WSN layout using a multi-objective genetic algorithm. In there formulation a fixed number of sensors had to be placed in order to maximize the coverage. In some applications most important is the network energy. In [16] is proposed ACO algorithm and in [17] is proposed evolutionary algorithm for this variant of the problem. The ACO algorithm in [18] taks in to account only the number of the sensors. In [19] are proposed several evolutionary algorithms to solve the problem. The genetic algorithm from [20] achieves similar solutions as the algorithms from [19], but it is tested on small test problems.

The number of sensor nodes should be kept low for economical reasons and the network needs to be connected. Finally the energy of the network is a key issue, that has to be taken into account, because the life time of the network depends on it.

7.2 Problem Formulation

A wireless sensor network consists of spatially distributed sensors which monitor the conditions of sensing area, such as temperature, sound, vibration, pressure, motion or pollutants [21, 22]. The development of wireless sensor networks was motivated by military applications such as surveillance and are now used in many industrial and civilian application areas, including industrial process monitoring and control, machine health monitoring, environment and habitat monitoring, health care applications, home automation, and traffic control [22].

A sensor node might vary in size from that of a box to the size of a grain of dust [22]. The cost of sensor nodes is similarly variable, ranging from hundreds of dollars to a few cents, depending on the complexity required of individual sensor nodes [22]. Size and cost constraints on sensor nodes result in corresponding constraints on resources such as energy, memory, computational speed and bandwidth [22].

A Wireless Sensor Network consists of sensor nodes. Each sensor node sense an area around itself called its sensing area. The sensing radius determines the sensitivity range of the sensor node and thus the sensing area. The communication radius determines how far the node can send his data. A special node in the WSN called High Energy Communication Node (HECN) is responsible for external access to the network. Therefore, every sensor node in the network must have communication with the HECN. Since the communication radius is often much smaller than the network size, direct links are not possible for peripheral nodes. A multi-hop communication path is then established for those nodes that do not have the HECN within their communication range. They transmit their date by other nodes which are closer to the HECN.

The WSN layout problem aims to decide the geographical position of the sensor nodes that form a WSN. In our formulation, sensor nodes has to be placed in a terrain providing full coverage with a minimal number of sensors and minimizing the energy spent in communications by any single node, while keeps the connectivity of the network. Minimal number of sensors means cheapest network for constructing. Minimal energy means cheapest network for exploitation. The energy of the network defines the lifetime of the network, how frequently the batteries need to be replaced. These are opposed objectives since the more nodes there are the lesser share of retransmissions they bear and in opposite, when we try to decrease the energy consumption normally we include nodes. Thus we look for a good balance between number of sensors and energy consumption.

In order to determine the energy spent by communications, the number of transmissions every node performs is calculated. The WSN operates by rounds: In a round, every node collects the data and sends it to the HECN. Every node transmits the information packets to the neighbor that is closest to the HECN, or the HECN itself if it is within the communication range. When several neighbors are tied for the shortest distance from the HECN, the traffic is distributed among them. That is, if a node has n neighbors tied for shortest distance from HECN, each one receives $1/n$ of its traffic load. Therefore, every node has a traffic load equal to 1 (corresponding to its own sent data) plus the sum of all traffic loads received from neighbors that are farther from the HECN. On one hand the sensing area need to be fully covered. On the other hand, the number of sensor nodes must be kept as low as possible, since using many nodes represents a high cost of the network, possibly influences of the environment. The objectives of this problem is to minimize network energy and the number of sensors deployed while the area is fully covered and connected.

7.3 ACO for Wireless Sensor Network Positioning

The WSN Layout problem is a hard combinatorial optimization problem which needs an exponential amount of computational resources (NP-hard). Exact methods and traditional numerical methods are unpractical for this kind of problems. Therefore normally the NP problems are solved with some metaheuristic method. Many of the existing solutions of WSN Layout problem come from the field of Evolutionary Computation [19, 23]. After analyzing them, we noticed that the ACO is based on solution construction, most of other metaheuristics are based on improving current solution. Therefore ACO is appropriate for problems with restrictive constraints.

In our implementation we use MAX-MIN Ant System (MMAS) [24], which is one of the more popular ant approaches. The main feature of MMAS is using a fixed upper bound τ_{max} and a lower bound τ_{min} of the pheromone trails. Thus the accumulation of big amounts of pheromone by part of the possible movements and repetition of same solutions is partially prevented on one side, and the amount of the pheromone to decrease a lot of and to become close to zero and the element to be unused is partially prevented in another side.

One of the crucial point of the ACO algorithms is construction of the graph of the problem. We need to chose which elements of the problem will correspond to the nodes and the meaning of the arcs, where is more appropriate to deposit the pheromone—on the nodes or on the arcs. In our implementation the WSN layout problem is represented by two graph, it is one of our contributions. The terrain is modeled by grid $G = \{g_{ij}\}_{N \times M}$, where M and N are the size of the sensing region. By the graph G we calculate the coverage of the terrain. We use another graph $G1_{N1 \times M1}$, on which nodes we map the sensors, where $N1 \leq N$ and $M1 \leq M$. The parameters $M1$ and $N1$ depend of the sensing and communication radius. Thus we decrease the number of calculations the algorithm performs, respectively the running time. The pheromone is related with location sites $Ph = \{ph_{ij}\}_{N1 \times M1}$, the initial pheromone can be a small value, for example $1/n_{ants}$. The point, where the HECN is located, is included in the solutions like first point (zero point).

Every ant starts to create the rest of the solution from a random node which communicates with central one, thus the different start of every ant in every iteration is guaranteed. The ant chooses the next position by the ACO probabilistic rule. It chooses the point having the highest probability. If there are more than one point with the same probability, the ant chooses one of them randomly.

The construction of heuristic information is another crucial points of the ACO algorithm. The heuristic information needs to be constructed thus, to manage the ants to look for better solutions and to avoid unfeasible solutions. For some kinds of problems it is not obvious how to prepare it. One needs to combine different elements of the problem to most appropriate way. The proposed by us heuristic information is a product of three parameters as follows:

$$\eta_{ij}(t) = s_{ij} l_{ij} (1 - b_{ij}), \tag{7.1}$$

where s_{ij} is the number of uncovered points, which the new sensor will cover, and

$$l_{ij} = \begin{cases} 1 \text{ if communication exists} \\ 0 \text{ if there is not communication} \end{cases} \tag{7.2}$$

b is the solution matrix and the matrix element $b_{ij} = 1$ when there is sensor on the node (i, j) of the graph $G1$, otherwise $b_{ij} = 0$. With s_{ij} we try to locally increase the covered points, more new covered points leads eventually to less number of sensors. With l_{ij} we guarantee that all sensors will be connected; with rule $(1 - b_{ij})$ we guarantee that the position is not chosen yet and no more than one sensor will be mapped on the same node of the graph $G1$. When $p_{ij} = 0$ for all values of i and j the search stops. Thus, the construction of the solution stops if no more free positions, or all points are covered or new communication is impossible.

Multi-objective optimization (MOP) is the process of simultaneously optimizing two or more conflicting objectives subject to certain constraints. If a multi-objective problem is well formed, there should not be a single solution that simultaneously minimizes each objective to its fullest.

Multi-Objective Optimization (MOP) has his roots in the nineteenth century in the work of Edgeworth and Pareto in economics [25]. The optimal solution for MOP is not a single solution as for mono-objective optimization problems, but a set of solutions defined as Pareto optimal solutions. A solution is Pareto optimal if it is not possible to improve a given objective without deteriorating at least another objective. The main goal of the resolution of a multi-objective problem is to obtain the Pareto optimal set and consequently the Pareto front. One solution dominates another if minimum one of its components is better than the same component of other solutions and other components are not worse. The Pareto front is the set of non dominated solutions.

7.4 Conversion of WSN Problem to Mono-Objective by Multiplication

The WSN layout problem is a NP-hard multi-objective problem. We simplify it converting it to mono-objective. Normally converting from multi to mono objective make worse the achieved solutions. Therefore is very important how the new objective function will be constructed. We propose the following objective function:

$$C(V_k) = f_1(V_k) \times f_2(V_k)\,, \tag{7.3}$$

where $f_1(V_k)$ is the number of the sensors achieved by the k-th ant and $f_2(V_k)$ is the energy of the solution of the k-th ant, V_k is the solution constructed by the k-th ant. Our goal is to minimize the both functions f_1 and f_2, which leads to the minimization of the new objective function C.

With our algorithm we can solve WSN layout problem on any rectangular area and the HECN can be fixed on any point on the area. The reported results are for an WSN problem instance where a terrain of 500×500 points has to be covered using sensors with coverage and communication radii covered 30 points (see Table 7.1). We choose this example because other authors use the same and to can compare achieved results. The terrain has an area of 250,000 points in total, and each sensor covers 2,827 points, meaning that in ideal conditions only 89 would be necessary. Now, these ideal conditions do not exist since they would imply that no overlap exists between any two nodes sensing areas, which is impossible due to their geometrical shape (circle). Therefore, the expected minimum number of nodes for full-coverage is higher than 89. Thus the graph G consists of 500×500 nodes. When we apply our ACO algorithm on this example the graph $G1$ consists of 50×50 nodes which are 100 times less than the graph G. The number of nodes of the graph $G1$ is proportional to the number of the nodes of the graph G and is proportional to the number of points covered by coverage and communication radius. Thus the nodes of the graph $G1$ are mapped on nodes of the graph G and coverage and communication radii cover 3 points from the graph $G1$. In our example the HECN is fixed in the center of the area.

An example of solution that achieves full coverage of the region is a square grid formed by the sensors separated by 30 points. Starting at the HECN, 250 points have to be covered to each side of the terrain, requiring 8 sensors. Therefore the grid has 17 (8 + 8 + 1) rows and 17 columns, thus 289 sensors including the HECN. In this symmetrical configuration there are four nodes directly connected to the HECN, so the complete traffic of the network 288 messages per round is evenly divided among them. This results in the most loaded nodes having a load of 72 messages. So this candidate solution obtains (288, 72). We will call this solution symmetric solution. One solution is considered good if it dominates the symmetric one. We apply MAX-MIN ant algorithm with the following parameters: $\alpha = \beta = 1$, $\rho = 0.5$. When the number of ants is double the running time is doubled. The number of used ants is from 1 to 10 and the maximum number of iterations is 60 (3 h per ant). We study the influence of the number of ants on the quality of the solutions.

Table 7.1 Problem parameters

Terrain	500×500
Coverage radius	30
Communication Radius	30

Table 7.2 Comparison with other metaheuristics

Algorithm	Min sensors	Min energy
Symmetric	(288,72)	(288,72)
MOEA	(260, 123)	(291,36)
NSGA-II	(262,83)	(277,41)
IBEA$_{HD}$	(265,83)	(275,41)
ACO	(225,63)	(237,51)

Table 7.3 Experimental results

Ants	Pareto front (Energy)														
1	53										58				63
2	53					54		55					57		
3		53			54			55					57		
4	52					53		55					57		
5	52					53	54	55					57		
6			51	53			54	55					57		
7			51			53	54						57		
8			51	53			54						57		
9			51	53			54						57		
10			51	53			54						57		
Sensor	239	238	237	236	235	234	233	232	231	230	229	228	227	226	225

In Table 7.2 we compare the results achieved by our ACO algorithm and by evolutionary algorithms [19] (with respect to the sensors and with respect to the energy). We observe that our mono-objective ACO algorithm achieves solutions with less number of sensors than evolutionary algorithms. The energy of the solution with less energy is a little bit more than this achieved by evolutionary algorithms, but the number of sensors is much less.

We run our ACO algorithm 30 times with every one of the ants number. We extract the Pareto front from the optimal solutions of this 30 runs. On the Table 7.3 we show the achieved nondominated solutions (Pareto fronts). Analyzing the table we observe that the Pareto front achieved by 6 ants dominates the Pareto fronts achieved by 1, 2, 3 and 4 ants. The Pareto fronts achieved by 8, 9 and 10 ants are identical. The Pareto Front achieved by 7 ants dominates this achieved by 8 ants and respectively by 9 and 10 ants. The Pareto front achieved by 4 ants is a part of the Pareto front achieved by 5 ants, but consists one solution less.

We prepare the Pareto front achieved by all runs of the algorithm with any number of ants (from 1 to 10) and we call it common Pareto front. Let we have a set of number of sensors from 225 to 239. If for some number of sensors there are not corresponding energy in the common Pareto front, we put the energy to be equal to the point of the front with less number of sensors. We can do

Table 7.4 Distances from Extended front

Ants	5	6	7
Distance	11	2	2

this because if we take some solution and if we include a sensor close to the HECN it will not increase the value of the energy and will increase with 1 only the number of the sensors. Thus there is corresponding energy to any number of nodes. This front we will call Extended front. In our case the Extended front is {(239, 51), (238,51), (237,51), (236,53), (235,53), (234,53), (233,54), (232,55), (231,57), (230,57), (229,57), (228,57), (227,57), (226,57), (225,57)}.

We include additional criteria to can decide which Pareto front is better. We will calculate the distance between a Pareto front and the Extended front. To calculate the distance we extend every of the Pareto fronts in a similar way as the common Pareto front. The distance between a Pareto front and the Extended front is the sum of the distances between the points with a same number of sensors, or it is the difference between their energy. These distances are always positive because the Extended front dominates the Pareto fronts. Thus by this criteria the best Pareto front will be the closest to the Extended front.

In Table 7.4 we show the distances between the Extended front and the Pareto fronts achieved by 5, 6 and 7 ants. Analyzing the Table 7.4 we conclude that the distance between the Extended front and the Pareto front achieved by 6 and 7 ants is shortest. When the ACO algorithm uses more ants, then the execution time increase and the used memory by the algorithm increases, which is very important when we solve large problems. Thus we decide that the best number of ants to solve wireless sensor layout problem is equal to 6.

7.5 Conversion of WSN Problem to Mono-Objective by Wheighted Sum

In this section the objective function is a sum of the number of sensors and the energy of the network and we search for solution which minimizes it. The new objective function is as follows:

$$C(V_k) = \frac{f_1(V_k)}{\max\limits_i f_1(V_i)} + \frac{f_2(V_k)}{\max\limits_i f_2(V_i)} \tag{7.4}$$

where V_k is the solution constructed by the ant k and $f_1(V_k)$ and $f_2(V_k)$ are the number of sensors and energy corresponding to the solution V_k. Dividing f_1 by $\max f_1$ and f_2 by $\max f_2$ is a sort to normalize the values of f_1 and f_2 respectively, the maximum is over the solutions from the first iteration. Thus when the energy and/or number

Table 7.5 Comparison with other metaheuristics

Algorithm	Min sensors	Min energy
Symmetric	(288,72)	(288,72)
MOEA	(260, 123)	(291,36)
NSGA-II	(262,83)	(277,41)
$IBEA_{HD}$	(265,83)	(275,41)
ACOprod	(225,63)	(237,51)
ACOsum	(226,58)	(234,48)

of sensors decreases the value of the objective function will decrease and the two components have equal influence.

We use the test problem from the previous section. After several runs of the algorithm we specify the most appropriate values of its parameters. We apply MAX-MIN ant algorithm with the following parameters: $\alpha = \beta = 1$, $\rho = 0.5$. When the number of ants is double the running time is doubled. The number of used ants is from 1 to 10 and the maximum number of iterations is 60 (3 h per ant).

In Table 7.5 are reported best found results (with respect to the sensors and with respect to the energy) achieved by several metaheuristic methods. We compare our ACO algorithm results with results obtained by the evolutionary algorithms in [19] and the symmetric solution. These evolutionary algorithms performs like multi-objective and reports non-dominated solutions. ACOprod is the ACO algorithm when the objective function is the product of the two objective functions and ACOsum is when the objective function is a sum of the normalized value of the two objective functions. The two variants of ACO find better solutions comparing with other algorithms. ACOprod finds solutions with less sensors than ACOsum, but ACOsum finds solutions with less energy.

We perform 30 independent runs of the ACOsum algorithm, because for statistical point of view the number of runs need to be minimum 30. The achieved numbers of sensors are in the interval [226, 247]. Thus the worst number of sensors achieved by ACOsum algorithm is less than the best number of sensors achieved by other mentioned algorithms. The energy is a little bit more than the evolutionary algorithms, but the number of sensors is much less.

We learn the influence of the number of ants on algorithm performance and quality of the achieved solutions. Our aim is to find optimal number of ants, first according the quality of the solutions and second—according the computational resources.

We run our ACO algorithm 30 times. We prepare Pareto front for over 30 runs for every ant number extracting the achieved non-dominated solutions. On the Table 7.6 we show the achieved non-dominated solutions (Pareto fronts). In the left column are the number of sensors and in other columns is the energy corresponding to this number of sensors and number of ants. We observe that the front achieved by 8 ants dominates the fronts achieved by 1, 2, 3, 4, 5, 6 and 7 ants. The fronts achieved by 8, 9 and 10 ants do not dominate each other. To

Table 7.6 Pareto fronts

Sensors	Ants									
	1	2	3	4	5	6	7	8	9	10
239	55			50						
238										
237	56									
236										
235		52	52	52					52	
234					51	51	51	48		51
233									54	54
232								51	55	55
231	58	58			56	56	56		51	
230		59	57	57	57	57	57	54	57	
229		60		60	60	60	58	56		
228	60									
227			60	61				57	58	58
226								58		

Table 7.7 Distances from Extended front

Ants	8	9	10
Distance	3	4	3

estimate them we prepare extended Pareto front similar to previous section. Thus there is corresponding energy to any number of nodes. In our case the extended front is $\{(226, 58), (227, 58), (228, 58), (229, 58), (230, 57), (231, 56), (232, 55), (233, 54), (234, 51), (235, 51)\}$. We calculate the distance between a Pareto front and the Extended front.

Regarding Table 7.7 we observe that the distance of the fronts achieved by 8 and 10 ants is 3 and the distance of the front achieved by 9 ants is 4. Thus the fronts achieved by 8 and 10 ants are with a equal quality and they are better than the front achieved by 9 ants. We consider that the algorithm performs better with 8 ants because the solution (226, 58) dominates solution (227, 58) and when the number of ants is less the used computational resources are less, running time is less and the used memory is less.

Other variant of the objective function is weighted sum. Than the objective function is as follows:

$$C(V_k) = \lambda \frac{f_1(V_k)}{\max_i f_1(V_i)} + (1 - \lambda) \frac{f_2(V_k)}{\max_i f_2(V_i)} \tag{7.5}$$

Table 7.8 Non-dominated solutions for weighted sum

λ	Nondominated solutions
0.125	(238,49) (235,51) (229,58) (227,60)
0.250	(238,49) (235,51) (228,58)
0.375	(238,49) (235,51) (229,57) (227,58) (226,60)
0.500	(238,49) (235,56) (229,57) (228,58) (226,60)
0.625	(238,49) (235,51) (229,57) (227,68) (226,60)
0.750	(238,49) (235,51) (229,57) (227,68) (226,60)
0.875	(238,49) (230,51) (225,57)

Table 7.9 Comparison with other metaheuristics

Algorithm	Min sensors	Min energy
Symmetric	(288,72)	(288,72)
MOEA	(260,123)	(291,36)
NSGA-II	(262,83)	(277,41)
$IBEA_{HD}$	(265,83)	(275,41)
ACOprod	(225,63)	(237,51)
ACOsum	(226,58)	(234,48)
ACOweight	(225,57)	(238,49)

The parameter λ has values from the set $\{0.125, 0.250, 0.375, 00.5, 0.625, 0.750, 0.875\}$ In Table 7.8 are reported non-dominated solutions for every value of the λ, and in Table 7.9 is reported comparison with other algorithms.

Our ACO algorithm with weighted sum (ACOweight) achieves best solutions when $\lambda = 0.875$ or when the number of sensors has more influence in the objective function than the energy. According solution with minimal number of sensors ACOweight achieves solution which dominates the solutions with minimal sensors achieved by ACOprod and ACOsum. According the minimal energy, the solution achieved by ACOsum dominates these achieved bu ACOprod and ACOweight.

7.6 Multi-objective WSN Problem

In this section we apply multi-objective ant colony optimization algorithm to solve the WSN problem, called ACOmulti. For pheromone updating we use fitness function instead of objective function. The role of the fitness function is to estimate the achieved solutions. The aim is to add more pheromone on non-dominated solutions and thus to force the ants to search around them for new non-dominated solutions. The fitness function is constructed as follows:

Table 7.10 Pareto fronts, example

Sensors	Ants									
	1	2	3	4	5	6	7	8	9	10
244					52					
243										
242										
241										
240	53	53								
239	56			50						
238			53							
237										
236										
235			54						50	
234					53			48	53	
233							51			
232			55	51	54	50	52	51		48
231		55			55		53			
230	57					52		54		
229	58			55				56		56
228									54	
227		57	57		57	56		57		
226	59	95	73	57	59	57	56			
225				58	60	58	57	58		57
224	61				88	65	61	59	57	71
223					89	81				

$$F(V_k) = \frac{f_1(V_k)}{\max_i f_1(V_i)} + \frac{f_2(V_k)}{\max_i f_2(V_i)} \tag{7.6}$$

or for fitness function we use the normalized sum of the objective functions.

We study the influence of the number of ants on the quality of the solutions. We fixed the number of the iterations to be 60 (3 h per ant) and the number of ants to have following values $\{1, 2, 3, 4, 5, 6, 7, 8, 9, 10\}$. We use the same test problem as in previous sections. We perform 30 independent runs of the ACOmulti algorithm and the achieved numbers of sensors are in the interval [223, 247]. Thus the worst number of sensors achieved by ACOmulti algorithm is less than the best number of sensors achieved by the mentioned evolutionary algorithms.

In Table 7.10 we show the achieved non dominated solutions (Pareto fronts) by ACOmulti. In the left column is the number of sensors and in other columns is the energy corresponding to this number of sensors and the number of ants. Analyzing the Table 7.10 we observe that the Pareto front achieved by 6 ants dominates the

Table 7.11 Distances from Extended front

Ants	6	7	8	9	10
Distance	20	23	21	22	29

Pareto fronts achieved by 1, 2, 3, 4 and 5 ants. The is not dominance between Pareto fronts achieved by 6, 7, 8, 9 and 10 ants and we cannot say which of them is better.

We prepare a Pareto front achieved by all runs of the algorithm with any number of ants (from 6 to 10) and we call it a common Pareto front. The common Pareto front is {(232,48), (230,52), (228,54), (226,56), (224,57), (223,81)}. Let us have a set of number of sensors from 223 to 244. If for some number of sensors there is not corresponding energy in the common Pareto front, we put the energy to be equal to the point of the front with lesser number of sensors. We can do this because, if we take some solution and if we include a sensor close to the HECN it will not increase the value of the energy and will increase by 1 only the number of the sensors. Thus, there is corresponding energy to any number of nodes. This front we will call the Extended front. The Extended front for our example is {(234,48), (233,48), (232,48), (231,52), (230,52), (229,54), (228,54), (227,56), (226,56), (225,57), (224,57), (223,81)}. We calculated the distance between a Pareto front and the Extended front. To calculate the distance, we extend every element of Pareto fronts in a similar way as the Extended front. The distance between a Pareto front and the Extended front is the sum of distances between the points with a same number of sensors, or it is the difference between their energy. These distances are always positive because the Extended front dominates the Pareto fronts. Thus, by this criteria, the best Pareto front will be the closest to the Extended front.

In Table 7.11 we show the distances between the Extended front and the Pareto fronts achieved by 6, 7, 8, 9 and 10 ants. Analyzing the Table 7.11 we conclude that the distance between the Extended front and the Pareto front achieved by 6 ants is the shortest. Thus, by our criteria, the Pareto front (solutions) achieved by 6 ants is better.

In Table 7.12 we compare our multi-objective ACO algorithm results with results obtained by the evolutionary algorithms in [19], the mono-objective ACO algorithms from previous sections and the symmetric solution.

We observe that the quality of the achieved solutions by multi-objective ACO are better than the quality achieved by mono-objective ACO algorithms. The solution with minimal number of sensors, which dominates other solutions with minimal number of sensors is (224, 57). ACOmulti achieves solution even with 223 sensors, but with very high value of energy, which do not dominates symmetric solution.

We also use the Hyper volume [26] as a quality indicator of achieved solutions. Mathematically, for each solution a hypercube is constructed with a reference point W and the solution as the diagonal corners of the hypercube. The reference point can be found simply by constructing a vector of worst objective function values. Thereafter, a

Table 7.12 Comparison with other metaheuristics

Algorithm	Min sensors	Min energy
Symmetric	(288,72)	(288,72)
MOEA	(260, 123)	(291,36)
NSGA-II	(262,83)	(277,41)
$IBEA_{HD}$	(265,83)	(275,41)
ACOprod	(225,63)	(237,51)
ACOsum	(226,58)	(234,48)
ACOweight	(225,57)	(238,49)
ACOmulti	(224,57)	(234,48)

Table 7.13 Solution hyper volume

Algorithm	Mean Hyper volume	Max Hyper volume	Min Hyper volume
MOEA	0.7388	0.7847	–
NSGA-II	0.7306	0.7868	–
$IBEA_{HD}$	0.7280	0.7704	–
ACOmulti	0.9440	0.9900	0.7769

union of all hyper cubes is found and its hyper volume (IHV) is calculated. Algorithms with larger IHV values are desirable.

Table 7.13 shows the mean hyper volume of the fronts obtained by the different algorithms. In multi-objective optimization there is no unique manner to present the data obtained from the executions. We employ the hyper volume metric in order to measure the solutions obtained by different algorithms. We observe that according the hyper volume the ACOmulti algorithm obtains much better results then other algorithms. The three genetic algorithms obtain statistically similar results. The minimal hyper volume obtained by ACOmulti algorithm is similar to the maximal hyper volume attained by others. Thus according hyper volume criterion our multi-objective ACO algorithm performs better than the evolutionary algorithms.

7.7 Different InterCriteria Analysis of Variants of ACO Algorithm

In the previous sections of this chapter are proposed different variants of ACO Algorithm to solve WSN problem: multi-objective ACO algorithm; mono-objective by multiplying the two objective functions; mono-objective by summing the two objective functions. We apply our algorithms on rectangular area consisting 500×500 points and the communication and coverage radius of every sensor cover 30 points. The number of used ants is from 1 to 10.

Table 7.14 Consonance and dissonance scale [28]

Interval of $\mu_{C,C'}$	Meaning
[0.00–0.05]	Strong negative consonance (SNC)
(0.05–0.15]	Negative consonance (NC)
(0.15–0.25]	Weak negative consonance (WNC)
(0.25–0.33]	Weak dissonance (WD)
(0.33–0.43]	Dissonance (D)
(0.43–0.57]	Strong dissonance (SD)
(0.57–0.67]	Dissonance (D)
(0.67–0.75]	Weak dissonance (WD)
(0.75–0.85]	Weak positive consonance (WPC)
(0.85–0.95]	Positive consonance (PC)
(0.95–1.00]	Strong positive consonance (SPC)

The input Index Matrix (IM) and the full set of obtained numerical results could be found in http://intercriteria.net/studies/aco/.

ICA is applied over the data presented in the input IM, where different ACO algorithms (*ACOu*—mono-objective with multiplication, *ACOs*—mono-objective with sum and *ACOm*—multi-objective) are presented as criteria and number of sensors [223, 224, 225, 226, 227, 228, 229, 230, 231, 232, 233, 234, 235, 236, 237, 238, 239, 240, 241, 242, 243, 244]—as objects. The cross-platform software for ICrA approach, ICrAData, is used [27].

The obtained ICrA results are analyzed based on the proposed in [28] consonance and dissonance scale. For ease of use the scheme for defining the consonance and dissonance between each pair of criteria is here presented in Table 7.14.

The obtained most significant and interesting ICrA results are presented in tables bellow.

7.7.1 Results from Application of μ-Biased ICrA

All results, based on application of μ-biased ICrA, are visualized on Fig. 7.1 within the specific triangular geometrical interpretation of IFSs, thus allowing us to order these results according simultaneously to the degrees of "agreement" $\mu_{C,C'}$ and "disagreement" $\nu_{C,C'}$ of the intuitionistic fuzzy pairs.

As it can be seen from Fig. 7.1 the ν-values for all considered pairs are zero ($\nu_{C,C'} = 0$). There are some results with $\pi_{C,C'}$ in the range between $\pi_{C,C'} = 0.02$ and $\pi_{C,C'} = 0.32$.

Table 7.15 displays the criteria pairs that are in SPC. Only the relations between algorithms with different objective functions are shown. There are 24 more criteria pairs that are also in SPC. These criteria pairs correspond to the algorithms with the

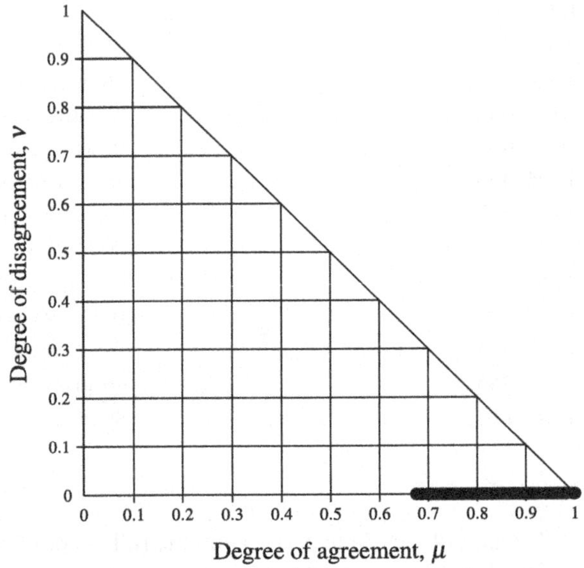

Fig. 7.1 Presentation of ICrA results in the intuitionistic fuzzy interpretation triangle—μ-biased ICrA

Table 7.15 Results from μ-biased ICrA—pairs in SPC

Criteria pairs of ACO algorithms	Value of $\mu_{C,C'}$
$ACOs8 - ACOm8$	0.98
$ACOs5 - ACOm8$	0.96
$ACOs6 - ACOm8$	0.96
$ACOs7 - ACOm8$	0.96
$ACOu3 - ACOm3$	0.95

same objective function and close number of ants, for example $ACOu8 - ACOu9$ or $ACOs2 - ACOs3$. As some exceptions, for the criteria pairs $ACOu1 - ACOu2$ and $ACOs4 - ACOs5$, a WPC is monitored and for the $ACOu5 - ACOu6$—WD. It should be noted that these pairs correspond to the case of mono-criteria variant of the problem. These pairs are not shown in the table. There is only one criteria pair corresponding to the multi-criteria case—$ACOm6 - ACOm10$.

The presented criteria pairs show that the SPC is observed for the four cases for $ACOs - ACOm$, and for one—$ACOu - ACOm$. There is not observed strong relation between results using $ACOu$ and $ACOs$. It could be summarized that SPC is only observed between mono-objective algorithms where the number of ants is nominally close, i.e. 5 and 6 ants or 7 and 8 ants. This observation is logical.

In Table 7.16 again only the results from different objective functions, that are in PC, are displayed. In this case the pairs of multi-criteria algorithms with a nominally close number of ants are appeared, as well as the mono-criteria with a bit larger difference between number of ants (e.g. $ACOs2 - ACOs10$, $ACOs3 - ACOs10$,

Table 7.16 Results from μ-biased ICrA—pairs in PC

Criteria pairs of ACO algorithms	Value of $\mu_{C,C'}$	Criteria pairs of ACO algorithms	Value of $\mu_{C,C'}$
$ACOu6 - ACOm3$	0.93	$ACOs8 - ACOm9$	0.88
$ACOs9 - ACOm8$	0.92	$ACOu2 - ACOm2$	0.88
$ACOu4 - ACOm3$	0.92	$ACOu4 - ACOs1$	0.88
$ACOs8 - ACOm5$	0.92	$ACOu6 - ACOs9$	0.88
$ACOu1 - ACOm1$	0.92	$ACOs9 - ACOm3$	0.87
$ACOu3 - ACOs4$	0.92	$ACOu2 - ACOm4$	0.87
$ACOu4 - ACOm4$	0.92	$ACOu5 - ACOs1$	0.87
$ACOu5 - ACOm3$	0.92	$ACOs9 - ACOm7$	0.87
$ACOu5 - ACOm4$	0.91	$ACOu6 - ACOs10$	0.87
$ACOu3 - ACOm4$	0.91	$ACOs7 - ACOm6$	0.86
$ACOs7 - ACOm5$	0.90	$ACOs8 - ACOm10$	0.86
$ACOs9 - ACOm9$	0.90	$ACOu7 - ACOs10$	0.86
$ACOu10 - ACOm3$	0.90	$ACOs7 - ACOm10$	0.86
$ACOu7 - ACOm3$	0.90	$ACOu10 - ACOs9$	0.86
$ACOu8 - ACOm3$	0.90	$ACOu5 - ACOs10$	0.86
$ACOu9 - ACOm3$	0.90	$ACOu7 - ACOs9$	0.86
$ACOs7 - ACOm7$	0.90	$ACOu8 - ACOs9$	0.86
$ACOu1 - ACOm4$	0.90	$ACOu9 - ACOs9$	0.86
$ACOs7 - ACOm9$	0.89	$ACOu1 - ACOs4$	0.85
$ACOs8 - ACOm7$	0.89	$ACOu2 - ACOm5$	0.85
$ACOu4 - ACOs4$	0.89	$ACOu2 - ACOs10$	0.85
$ACOu2 - ACOm3$	0.89	$ACOu4 - ACOs10$	0.85
$ACOu3 - ACOs1$	0.89	$ACOs9 - ACOm6$	0.85
$ACOu5 - ACOs4$	0.89	$ACOu2 - ACOs4$	0.85
$ACOs8 - ACOm6$	0.88	$ACOu3 - ACOs9$	0.85
$ACOs9 - ACOm5$	0.88	$ACOu6 - ACOs1$	0.85

$ACOu2 - ACOu10, ACOu7 - ACOu10$, etc.). The full set of the obtained numerical results could be found at http://intercriteria.net/studies/aco/.

From total 84 cases of criteria pairs between different algorithms 19 pairs are between $ACOs$ and $ACOu$ (mono-criteria) and the rest are between mono and multi-criteria ACO algorithms. That means that the relation between mono and multi-criteria algorithms are weaker than the relations between different mono-criteria algorithms. The last ones are mainly in the SPC.

The presented in Table 7.16 results show that the algorithms $ACOs - ACOu$ are more similar to each other than the $ACOu - ACOm$ algorithms.

Table 7.17 Results from μ-biased ICrA—pairs in WPC

Criteria pairs of ACO algorithms of ACO algorithms	Value of $\mu_{C,C'}$
$ACOs4 - ACOs10$	0.77
$ACOs4 - ACOs5$	0.82
$ACOs4 - ACOs6$	0.82
$ACOs4 - ACOs7$	0.82
$ACOs4 - ACOs8$	0.81
$ACOs4 - ACOs9$	0.84
$ACOu1 - ACOu2$	0.78
$ACOu1 - ACOu3$	0.81
$ACOu1 - ACOu4$	0.82
$ACOu1 - ACOu5$	0.82

163 criteria pairs are in WPC. 157 from them are between mono and multi-criteria ACO algorithms. Among the results only 6 pairs are between $ACOs - ACOs$ and 4 are between $ACOu - ACOu$. The results show that there os a weaker relation between mono and multi-criteria ACO algorithms. 137 of the criteria pair are between $ACOu - ACOs$ and $ACOu - ACOm$, i.e. $ACOu$ performance is more different in comparison with the performance of the $ACOs$ and $ACOm$ algorithms.

64 criteria pairs are in WD. Among them there are not criteria pairs between $ACOs - ACOs$ algorithms.

In Table 7.17 the criteria pairs between the same objective function that are in weak positive consonance are presented.

Considering the relations between the same objective function the influence of the different number of ants on the ACO performance is investigated. The results show that $ACOs$ are less sensitive to the number of ants, compared to the $ACOu$ and $ACOm$. The larger difference of the performance are observed for $ACOu$ and $ACOm$—53% of all criteria pairs. In 14% of the cases the WD is observed for the algorithms $ACOu - ACOu$ and $ACOm - ACOm$ with bigger difference between the number of ants (more than three), **which that** $ACOu$ and $ACOm$ highly influenced of the number of ants.

7.7.2 Results from Application of ν-Biased ICrA

All results, based on application of ν-biased ICrA, are visualized on Fig. 7.2 within the specific triangular geometrical interpretation of IFSs.

The presented results show that in this case the degree of uncertainty ($\nu_{C,C'}$) are not zero. The observed $\nu_{C,C'}$-values are further included in the tables with presented results.

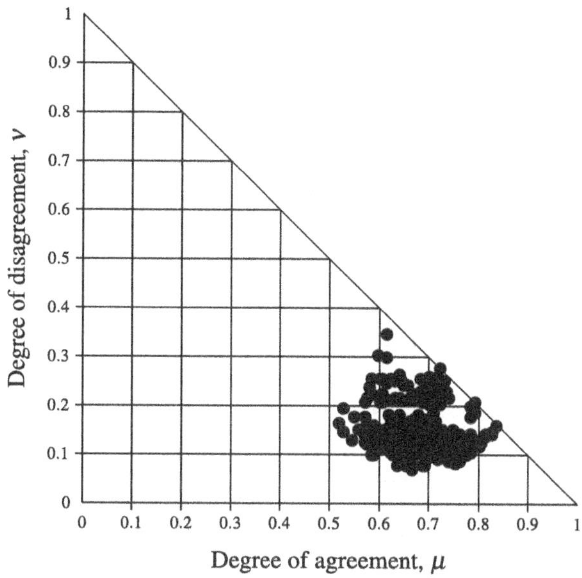

Fig. 7.2 Presentation of ICrA results in the intuitionistic fuzzy interpretation triangle—ν-biased ICrA

Table 7.18 Results from ν-biased ICrA—pairs in WPC and WD

Criteria pairs of ACO algorithms of ACO algorithms	Value of $\mu_{C,C'}$	Value of $\nu_{C,C'}$
$ACOs8 - ACOm8$	0.83	0.25
$ACOs5 - ACOm8$	0.71	0.25
$ACOs6 - ACOm8$	0.71	0.25
$ACOs7 - ACOm8$	0.71	0.25
$ACOu3 - ACOm3$	0.81	0.14

In Table 7.18 the results for criteria pairs, considered in the case of μ-biased ICrA, are displayed. In comparison with the results in subsection 7.7.1, the regarded criteria pairs are not in SPC. There are pairs in WD ($ACOs5 - ACOm8$, $ACOs6 - ACOm8$ and $ACOs7 - ACOm8$) or pairs in WPC ($ACOs8 - ACOm8$ and $ACOu3 - ACOm3$).

The presented results show that in case of ν-biased ICrA lower criteria correlations are observed compared to the results based on μ-biased ICrA. The main reason of this is the way of interpretation of equal results in the input index matrix (see http://intercriteria.net/studies/aco/). The equal results are assigned to the degree of "disagreement" and thus the value of $\nu_{C,C'}$ is increased. In a result, the degree $\mu_{C,C'}$ is decreased.

In Table 7.19 the results from different objective functions, considered in subsection 7.7.1, are presented with the corresponding $\nu_{C,C'}$-values.

Table 7.19 Results from ν-biased ICrA—pairs in WPC, WD and D

Criteria pairs of ACO algorithms	Value of $\mu_{C,C'}$	Value of $\nu_{C,C'}$	Criteria pairs of ACO algorithms	Value of $\mu_{C,C'}$	Value of $\nu_{C,C'}$
$ACOu6 - ACOm3$	**0.78**	**0.14**	$ACOs8 - ACOm9$	0.67	0.21
$ACOs9 - ACOm8$	0.72	0.20	$ACOu2 - ACOm2$	0.73	0.15
$ACOu4 - ACOm3$	**0.79**	**0.13**	$ACOu4 - ACOs1$	**0.77**	**0.10**
$ACOs8 - ACOm5$	0.71	0.20	$ACOu6 - ACOs9$	0.73	0.15
$ACOu1 - ACOm1$	0.71	0.21	$ACOs9 - ACOm3$	0.74	0.14
$ACOu3 - ACOs4$	**0.80**	**0.12**	$ACOu2 - ACOm4$	**0.76**	**0.11**
$ACOu4 - ACOm4$	**0.78**	**0.13**	$ACOu5 - ACOs1$	**0.77**	**0.10**
$ACOu5 - ACOm3$	0.73	0.11	$ACOs9 - ACOm7$	0.65	0.21
$ACOu5 - ACOm4$	**0.78**	**0.13**	$ACOu6 - ACOs10$	0.69	0.17
$ACOu3 - ACOm4$	**0.79**	**0.13**	$ACOs7 - ACOm6$	0.61	0.25
$ACOs7 - ACOm5$	0.70	0.20	$ACOs8 - ACOm10$	0.61	0.25
$ACOs9 - ACOm9$	0.70	0.21	$ACOu7 - ACOs10$	0.68	0.18
$ACOu10 - ACOm3$	**0.76**	**0.14**	$ACOs7 - ACOm10$	0.60	0.26
$ACOu7 - ACOm3$	**0.76**	**0.14**	$ACOu10 - ACOs9$	0.71	0.15
$ACOu8 - ACOm3$	**0.76**	**0.14**	$ACOu5 - ACOs10$	0.70	0.16
$ACOu9 - ACOm3$	**0.76**	**0.14**	$ACOu7 - ACOs9$	0.71	0.15
$ACOs7 - ACOm7$	0.65	0.25	$ACOu8 - ACOs9$	0.71	0.15
$ACOu1 - ACOm4$	0.71	0.19	$ACOu9 - ACOs9$	0.71	0.15
$ACOs7 - ACOm9$	0.67	0.23	$ACOu1 - ACOs4$	0.70	0.15
$ACOs8 - ACOm7$	0.65	0.24	$ACOu2 - ACOm5$	0.73	0.12
$ACOu4 - ACOs4$	**0.78**	**0.11**	$ACOu2 - ACOs10$	0.70	0.16
$ACOu2 - ACOm3$	**0.77**	**0.11**	$ACOu4 - ACOs10$	0.70	0.16
$ACOu3 - ACOs1$	**0.78**	**0.10**	$ACOs9 - ACOm6$	0.63	0.22
$ACOu5 - ACOs4$	**0.78**	**0.10**	$ACOu2 - ACOs4$	**0.76**	**0.09**
$ACOs8 - ACOm6$	0.63	0.25	$ACOu3 - ACOs9$	0.73	0.12
$ACOs9 - ACOm5$	0.71	0.17	$ACOu6 - ACOs1$	0.75	0.10

The criteria pairs that are in weak positive consonance are presented in bold. As it can be seen only 18 criteria pairs are in weak positive consonance. The rest of the criteria pairs are in weak dissonance or in dissonance.

In Table 7.20 the criteria pairs considered in the case of μ-biased ICrA as pairs in WPC are presented. In the case of ν-biased ICrA these criteria pairs are in weak dissonance and in dissonance (for criteria pairs $ACOs4 - ACOs10$ and $ACOu1 - ACOu2$).

In Table 7.21 only the criteria pairs between the same objective function that are in weak positive consonance (ν-biased ICrA) are presented. Although, the lower $\mu_{C,C'}$-values the results show that the performance of mono-objective ACO algorithms with sum ($ACOs$) are less sensitive to the number of ants, compared to the mono-objective ACO with multiplication ($ACOu$) and multi-objective ACO ($ACOm$).

Table 7.20 Results from ν-biased ICrA—pairs in WD and D

Criteria pairs of ACO algorithms	Value of $\mu_{C,C'}$	Value of $\nu_{C,C'}$
$ACOs4 - ACOs10$	0.66	0.11
$ACOs4 - ACOs5$	0.69	0.13
$ACOs4 - ACOs6$	0.69	0.13
$ACOs4 - ACOs7$	0.69	0.13
$ACOs4 - ACOs8$	0.69	0.12
$ACOs4 - ACOs9$	0.73	0.12
$ACOu1 - ACOu2$	0.66	0.12
$ACOu1 - ACOu3$	0.68	0.13
$ACOu1 - ACOu4$	0.68	0.14
$ACOu1 - ACOu5$	0.68	0.13

Table 7.21 Results from ν-biased ICrA—pairs in WPC

Criteria pairs of ACO algorithms	Value of $\mu_{C,C'}$	Value of $\nu_{C,C'}$
$ACOu4 - ACOu5$	0.84	0.16
$ACOu3 - ACOu4$	0.83	0.15
$ACOu3 - ACOu5$	0.83	0.14
$ACOu2 - ACOu4$	0.81	0.14
$ACOu2 - ACOu5$	0.81	0.14
$ACOu3 - ACOm3$	0.81	0.14
$ACOu2 - ACOu3$	0.81	0.13

7.7.3 Results from Application of Balanced ICrA

All results, based on application of Balanced ICrA, are visualized on Fig. 7.3 within the specific triangular geometrical interpretation of IFSs.

In comparison with the results from ν-biased, here the correlation between the criteria pairs preserve the $\mu_{C,C'}$-values. The $\nu_{C,C'}$-values are redistributed between the $\pi_{C,C'}$-values and $\nu_{C,C'}$-values.

Again the obtained in this case results are compared to the results observed in subsection 7.7.1. Table 7.22 displays the $\mu_{C,C'}$ and $\nu_{C,C'}$-values of the criteria pairs considered in case of μ-biased ICrA. In comparison with the results in subsection 7.7.1, the regarded criteria pairs are not in SPC, there are in WD ($ACOs5 - ACOm8$, $ACOs6 - ACOm8$ and $ACOs7 - ACOm8$) or in WPC ($ACOs8 - ACOm8$ and $ACOu3 - ACOm3$).

In Table 7.23 the results from different objective functions, considered in subsection 7.7.1, are displayed with the corresponding $\nu_{C,C'}$-values.

Table 7.22 Results from Balanced ICrA—pairs in WPC and WD

Criteria pairs of ACO algorithms	Value of $\mu_{C,C'}$	Value of $\nu_{C,C'}$
$ACOs8 - ACOm8$	0.86	0.13
$ACOs5 - ACOm8$	0.84	0.12
$ACOs6 - ACOm8$	0.84	0.12
$ACOs7 - ACOm8$	0.84	0.12
$ACOu3 - ACOm3$	0.88	0.07

Table 7.23 Results from Balanced ICrA—pairs in WPC, PC and WD

Criteria pairs of ACO algorithms	Value of $\mu_{C,C'}$	Value of $\nu_{C,C'}$	Criteria pairs of ACO algorithms	Value of $\mu_{C,C'}$	Value of $\nu_{C,C'}$
$ACOu6 - ACOm3$	**0.85**	**0.07**	$ACOs8 - ACOm9$	0.67	0.21
$ACOs9 - ACOm8$	**0.76**	**0.07**	$ACOu2 - ACOm2$	0.73	0.15
$ACOu4 - ACOm3$	**0.82**	**0.10**	$ACOu4 - ACOs1$	**0.83**	**0.05**
$ACOs8 - ACOm5$	**0.82**	**0.10**	$ACOu6 - ACOs9$	**0.81**	**0.07**
$ACOu1 - ACOm1$	**0.81**	**0.11**	$ACOs9 - ACOm3$	**0.81**	**0.07**
$ACOu3 - ACOs4$	**0.86**	**0.06**	$ACOu2 - ACOm4$	**0.82**	**0.06**
$ACOu4 - ACOm4$	**0.85**	**0.07**	$ACOu5 - ACOs1$	**0.82**	**0.05**
$ACOu5 - ACOm3$	**0.85**	**0.06**	$ACOs9 - ACOm7$	**0.76**	**0.11**
$ACOu5 - ACOm4$	**0.85**	**0.06**	$ACOu6 - ACOs10$	**0.78**	**0.09**
$ACOu3 - ACOm4$	**0.85**	**0.06**	$ACOs7 - ACOm6$	0.74	0.12
$ACOs7 - ACOm5$	**0.80**	**0.10**	$ACOs8 - ACOm10$	0.73	0.13
$ACOs9 - ACOm9$	**0.80**	**0.10**	$ACOu7 - ACOs10$	**0.77**	**0.09**
$ACOu10 - ACOm3$	**0.83**	**0.07**	$ACOs7 - ACOm10$	0.73	0.13
$ACOu7 - ACOm3$	**0.83**	**0.07**	$ACOu10 - ACOs9$	**0.78**	**0.07**
$ACOu8 - ACOm3$	**0.83**	**0.07**	$ACOu5 - ACOs10$	**0.78**	**0.08**
$ACOu9 - ACOm3$	**0.83**	**0.07**	$ACOu7 - ACOs9$	**0.78**	**0.07**
$ACOs7 - ACOm7$	**0.77**	**0.13**	$ACOu8 - ACOs9$	**0.78**	**0.07**
$ACOu1 - ACOm4$	**0.80**	**0.09**	$ACOu9 - ACOs9$	**0.78**	**0.07**
$ACOs7 - ACOm9$	**0.78**	**0.11**	$ACOu1 - ACOs4$	**0.78**	**0.08**
$ACOs8 - ACOm7$	**0.77**	**0.12**	$ACOu2 - ACOm5$	**0.79**	**0.06**
$ACOu4 - ACOs4$	**0.84**	**0.05**	$ACOu2 - ACOs10$	**0.77**	**0.08**
$ACOu2 - ACOm3$	**0.83**	**0.06**	$ACOu4 - ACOs10$	**0.77**	**0.08**
$ACOu3 - ACOs1$	**0.84**	**0.05**	$ACOs9 - ACOm6$	0.74	0.11
$ACOu5 - ACOs4$	**0.84**	**0.05**	$ACOu2 - ACOs4$	**0.81**	**0.04**
$ACOs8 - ACOm6$	**0.76**	**0.13**	$ACOu3 - ACOs9$	**0.79**	**0.06**
$ACOs9 - ACOm5$	**0.80**	**0.08**	$ACOu6 - ACOs1$	**0.80**	**0.05**

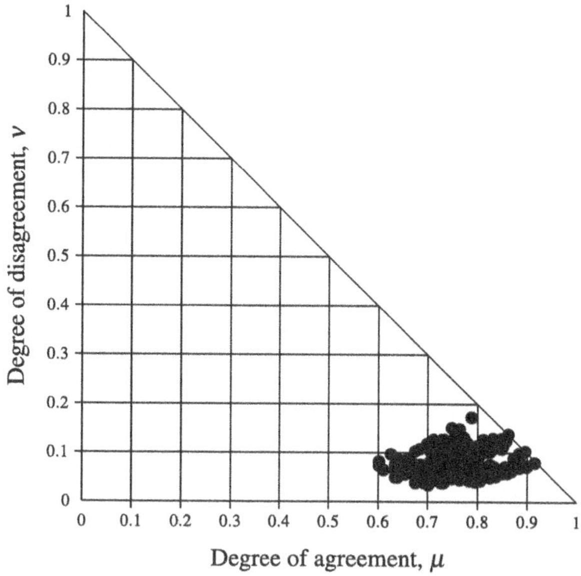

Fig. 7.3 Presentation of ICrA results in the intuitionistic fuzzy interpretation triangle—Balanced ICrA

The criteria pairs that are in weak positive consonance and in positive consonance are presented in bold. Compared to the case of ν-biased ICrA with 34 criteria pairs in weak dissonance and dissonance, here only 6 pairs are in weak dissonance. In case of μ-biased ICrA all listed in Table 7.23 criteria pairs are in positive consonance. So, Balanced ICrA shows performance similar to μ-biased ICrA, but still takes account of the repetitive numerical results. Thus, the $\pi_{C,C'}$-values variate from $\pi_{C,C'} = 0.02$ to $\pi_{C,C'} = 0.33$.

In Table 7.24 the criteria pairs considered in the case of μ-biased ICrA as pairs in WPC are presented. In the case of Balanced ICrA these criteria pairs are in weak dissonance with the exception of three pairs, namely $ACOs4 - ACOs5$, $ACOs4 - ACOs6$ and $ACOs4 - ACOs7$. These criteria pairs are in weak positive consonance. In comparison, in case of μ-biased ICrA, all pairs listed in Table 7.24 are in weak positive consonance. Applying Balanced ICrA algorithm it is shown that only $ACOs$ algorithm keeps the positive consonance between some of the criteria pairs. Thus, the conclusion that $ACOs$ algorithms are less sensitive to the number of ants, compared to the $ACOu$ and $ACOm$ algorithms, is confirmed under these more stringent conditions.

In the case of Balanced ICrA algorithm there are some criteria pairs between the same objective function that are in positive consonance. The results are presented in Table 7.25.

In the case of ν-biased ICrA there criteria pairs are in WPC and in the case of μ-biased ICrA they are in PC, even in SPC. If the correlation between given criteria pair (two ACO algorithms) is retained during application of different ICrA approaches,

Table 7.24 Results from Balanced ICrA—pairs in WD and WPC

Criteria pairs of ACO algorithms	Value of $\mu_{C,C'}$	Value of $\nu_{C,C'}$
$ACOs4 - ACOs10$	0.71	0.06
$ACOs4 - ACOs5$	**0.76**	**0.06**
$ACOs4 - ACOs6$	**0.76**	**0.06**
$ACOs4 - ACOs7$	**0.76**	**0.06**
$ACOs4 - ACOs8$	0.75	0.06
$ACOs4 - ACOs9$	0.79	0.06
$ACOu1 - ACOu2$	0.72	0.06
$ACOu1 - ACOu3$	0.75	0.06
$ACOu1 - ACOu4$	0.75	0.07
$ACOu1 - ACOu5$	0.75	0.07

Table 7.25 Results from Balanced ICrA—pairs in PC

Criteria pairs of ACO algorithms	Value of $\mu_{C,C'}$	Value of $\nu_{C,C'}$
$ACOu4 - ACOu5$	0.92	0.08
$ACOu3 - ACOu4$	0.90	0.07
$ACOu3 - ACOu5$	0.90	0.07
$ACOu2 - ACOu4$	0.89	0.07
$ACOu2 - ACOu5$	0.88	0.07
$ACOu3 - ACOm3$	0.88	0.07
$ACOu2 - ACOu3$	0.87	0.06

this means that it is really strong dependence of the pair. If the correlation is lost, it means that the resulting numerical results can lead us to an existing relationship.

7.7.4 *Results from Application of Unbiased ICrA*

All results, based on application of Unbiased ICrA, are visualized on Fig. 7.4 within the specific triangular geometrical interpretation of IFSs.

In this case the obtained $\mu_{C,C'}$-values are the same of those obtained in the case of ν-biased ICrA (subsection 7.7.2). The difference in the results is in the estimates of $\nu_{C,C'}$-values. All $\nu_{C,C'}$-values are zero, resulting in non zero degree of uncertainty $\pi_{C,C'}$. The $\pi_{C,C'}$-values variate from $\pi_{C,C'} = 0.16$ to $\pi_{C,C'} = 0.48$. The algorithm of Unbiased ICrA in case of repetitive numerical results in input index matrix increases the value of the degree of uncertainty $\pi_{C,C'}$. That is the reason of zero estimates for $\nu_{C,C'}$-values.

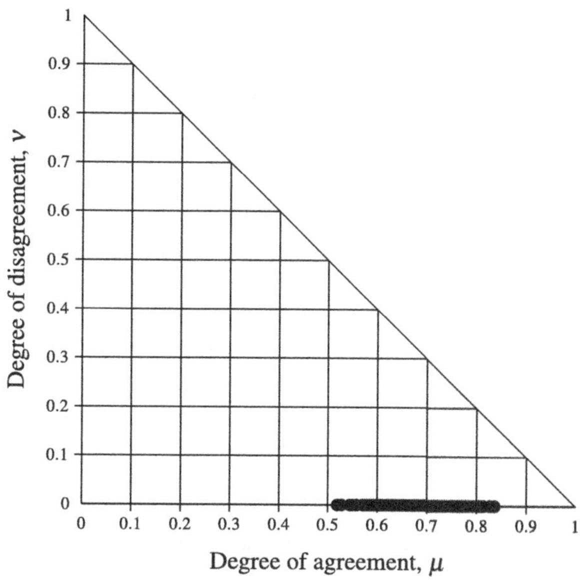

Fig. 7.4 Presentation of ICrA results in the intuitionistic fuzzy interpretation triangle—Unbiased ICrA

Table 7.26 Results from μ-biased and Unbiased ICrA

Criteria pairs	ν-biased ICrA			Unbiased ICrA		
of ACO algorithms	$\mu_{C,C'}$	$\nu_{C,C'}$	$\pi_{C,C'}$	$\mu_{C,C'}$	$\nu_{C,C'}$	$\pi_{C,C'}$
$ACOs8 - ACOm8$	0.83	0.25	0.08	0.83	0.00	0.17
$ACOs5 - ACOm8$	0.71	0.25	0.04	0.71	0.00	0.29
$ACOs6 - ACOm8$	0.71	0.25	0.04	0.71	0.00	0.29
$ACOs7 - ACOm8$	0.71	0.25	0.04	0.71	0.00	0.29
$ACOu3 - ACOm3$	0.81	0.14	0.05	0.81	0.00	0.19

The comparison of the here obtained results for different objective functions (criteria pairs in SPC in the case of μ-biased ICrA), with the results in the case of ν-biased ICrA algorithm, is presented in Table 7.26.

As it be can see the difference in the criteria pairs correlations is only in the interpretation of repetitive numerical results in the input matrix. In the case of μ-biased ICrA algorithm the repetitive numerical results are interpreted as degree of "disagreement", and in the case of Unbiased ICrA algorithm—as degree of "uncertainty".

7.8 Conclusion

In this chapter a multi-objective as well as several variants of mono-objective ACO algorithms are proposed to solve Wireless Sensor Network layout problem. The problem consists in providing the number and locations of the sensor nodes that form a WSN with full coverage of the sensing field and minimal number of sensors and energy. We compare achieved results with the best found results for this problem in a literature. We show that proposed ACO algorithms performs better than existing ones. The mono-objective algorithm with weighted sum achieves similar solutions as multi-objective ACO algorithm. We study the best value of the weights and found that it is better when the number of sensors has more influence in the objective function that the energy. We study the influence of the number of ants on the performance of the ACO algorithm, applied to the wireless sensor network. Smaller number of ants leads to the shorter running time and minimizes memory use, which is important for complex/large cases. We varied the number of ants, while fixing the number of iterations. Furthermore, we included the concept of an Extended front, as an additional tool to compare Pareto fronts that do not dominate each other.

The InterCriteria analysis is a powerful tool for studying relations between different objects. We study three variants of ACO algorithm applied on WSN problem. Every variant is tested with various number of ants, between 1 and 10. We search the correlation between variants of ACO and number of ants. WSN problem is a multi-objective problem. When it is converted to mono-objective by summing the two objective functions, the algorithm is less sensitive to the number of used ants. When the problem is solved like multi-objective, we observe bigger difference of algorithm performance according to the number of ants. There is greater similarity between performance of the two mono-objective variants, than between some of mono-objective and multi-objective variants. The InterCriteria analysis is applied in order to examine the relations between regarded hybrid ACO algorithms. The influence of the number of ants is analysed. Four different algorithms of ICrA—μ-biased, Balanced, ν-biased and Unbiased—are applied. The obtained results show that in case of correlation between given hybrid ACO algorithms this dependence is appeared regardless of the ICrA algorithm we use.

References

1. Fidanova, S., Marinov, P.: Optimal wireless sensor network coverage with ant colony optimization. In: Proceeding of International Conference on Swarm Intelligence. Paris, France (2011). Available via http://csi11.eisti.fr/papers/
2. Fidanova, S., Marinov, P.: Influence of the number of ants on mono-objective ant colony optimization algorithm for wireless sensor network layout. In: Proceeding of BGSIAM'12, pp. 59–66. Sofia, Bulgaria (2012)
3. Fidanova, S., Shindarov, M., Marinov, P.: Mono-objective algorithm for wireless sensor layout. In: Proceeding of OMCO-NET Conference, pp. 57–63. Southempton, UK (2012)

4. Fidanova, S., Shindarov, M., Marinov, P.: Multi-objective ant algorithm for wireless sensor network positioning. Proc. Bulgarian Acad. Sci. **66**(3), 353–360 (2013)
5. Fidanova, S., Marinov, P.: Number of ants versus number of iterations on ant colony optimization algorithm for wireless sensor layout. In: Proceeding of Robotics Automation and Mechatronics, pp. 90–93. Bankya, Bulgaria (2013)
6. Fidanova, S., Marinov, P., Paprzycki, M.: Influence of the number of ants on multy-objective ant colony optimization algorithm for wireless sensor network layout: large-scale scientific computing. Lect. Notes Comput. Sci. **8353**, 208–215. Springer (2014)
7. Fidanova, S., Marinov, P., Paprzycki, M.: Multi-objective ACO algorithm for WSN layout: performance according number of ants. J. Metaheuristics **3**(2), 149–161 (2014)
8. Shindarov, M., Fidanova, S., Marinov, P.: Wireless sensor positioning algorithm. In: Proceeding of IEEE Conference on Intelligent Systems, pp. 419–424. Sofia, Bulgaria (2012)
9. Deb, K., Pratap, A., Agrawal, S., Meyarivan, T.: A fast and elitist multiobjective genetic algorithm: Nsga-ii. IEEE Trans. Evol. Comput. **6**(2), 182–197 (2002)
10. Werner-Allen, G., Lorinez, K., Welsh, M., Marcillo, O., Jonson, J., Ruiz, M., Lees J.: Deploying a wireless sensor network on an active volcano. IEEE Internet Comput. **10**(2), 18–25 (2006)
11. Yuce, M.R., Ng, S.W., Myo, N.L., Khan, J.Y., Liu, W.: Wireless body sensor network using medical implant band. Med. Syst. **31**(6), 467–474 (2007)
12. Paek, J., Kothari, N., Chintalapudi, K., Rangwala, S., Govindan, R.: The performance of a wireless sensor network for structural health monitoring. In: Proceeding of 2nd European Workshop on Wireless Sensor Networks. Istanbul, Turkey, Jan 31–Feb 2 (2004). https://escholarship.org/uc/item/9bf7f3n5
13. Pottie, G.J., Kaiser, W.J.: Embedding the internet: wireless integrated network sensors. Commun. ACM **43**(5), 51–58 (2000)
14. Xu, Y., Heidemann, J., Estrin, D.: Geography informed energy conservation for ad hoc routing. Proceedings of the 7th ACM/IEEE Annual International Conference on Mobile Computing and Networking, Italy, pp. 70–84 (2001)
15. Jourdan, D.B.: Wireless sensor network planing with application to UWB localization in GPS-denied environments. PhD thesis, Massachusetts Institute of Technology (2000)
16. Hernandez, H., Blum, C.: Minimum energy broadcasting in wireless sensor networks: An ant colony optimization approach for a realistic antenna model. J. Appl. Soft Comput. **11**(8), 5684–5694 (2011)
17. Wolf, S., Mezz, P.: Evolutionary local search for the minimum energy broadcast problem. In: Cotta, C., van Hemezl, J. (eds.) VOCOP 2008. Lecture Notes in Computer Sciences, vol. 4972, pp. 61–72. Springer, Germany (2008)
18. Fidanova, S., Marinov, P., Alba, E.: ACO for optimal sensor layout. In: Filipe, J., Kacprzyk, J. (eds.) Proceeding of International Conference on Evolutionary Computing, pp. 5–9. Valencia, Spain. SciTePress-Science and Technology Publications Portugal. ISBN 978-989-8425-31-7 (2010)
19. Molina, G., Alba, E.: Talbi El-G: optimal sensor network layout using multi-objective metaheuristics. Univ. Comput. Sci. **14**(15), 2549–2565 (2008)
20. Konstantinidis, A., Yang, K., Zhang, Q., Zainalipour-Yazti, D.: A multi-objective evolutionary algorithm for the deployment and power assignment problem in wireless sensor networks. J. Comput. Netw. **54**(6), 960–976 (2010)
21. Akuildiz, I.F., Su, W., Sankarasubramaniam, Y., Cayrci, E.: Wireless sensor networks: a survey. Comput. Netw. **38**(4), 393–422. Elsevier (2001)
22. Romer, K., Mattern, F.: The design space of wireless sensor networks. IEEE Wireless Commun. **11**(6), 54–61 (2004)
23. Alba, E., Molina, G.: Optimal wireless sensor layout with metaheuristics: solving a large scale instance, large-scale scientific computing. Lect. Notes Comput. Sci. **4818**, 527–535. Springer (2008)
24. Shmygelska, A., Hoos, H.H.: An ant colony optimization algorithm for the 2D and 3D hydrophobic polar protein folding problem. BMC Bioinf. **6**(30) (2005). https://doi.org/10.1186/1471-2105-6-30

25. Mathur, V.K.: How well do we know Pareto optimality? J. Econ. Educ. **22**(2), 172–178 (1991)
26. Saleh, H.A., Dare, P.J.: Heuristic methods for designing Global Positioning System Surveying Network in the Republic of Seychelles. Arab. J. Sci. Eng. **26**, 73–93 (2002)
27. Ikonomov, N., Vassilev, P., Roeva, O.: ICrAData software for InterCriteria analysis. Inter. J. Bioautomation **22**(2) (2018)
28. Atanassov, K., Atanassova, V., Gluhchev, G.: InterCriteria analysis: ideas and problems. Notes Intuitionistic Fuzzy Sets **21**(1), 81–88 (2015)

Chapter 8
ACO for Image Edge Detection

The text of this chapter is based on [1]. The aim of the image edge detection is to find the points, in a digital image, at which the brightness level changes sharply. Normally they are curved lines called edges. Edge detection is a fundamental tool in image processing, machine vision and computer vision, particularly in the areas of feature detection and feature extraction. Edge detection may lead to finding the boundaries of objects. It is one of the fundamental steps in image analysis. Edge detection is a hard computational problem. In this paper we apply a multiagent system. The idea comes from ant colony optimization. We use the swarm intelligence of the ants to search the image edges.

8.1 Problem Formulation

Edge detection plays an important role in image processing. It is a main stage in image segmentation, pattern recognition and scene analysis [2–4]. Edge detection refers to the process of identifying and locating sharp discontinuities in an image. The discontinuities are abrupt changes in pixel intensity which characterize boundaries of objects and textures in a scene. Classical methods of edge detection involve convolving the image with an operator, a 2-D filter, which is constructed to be sensitive to large gradients in the image while at the same time returning values of zero in uniform regions. There is an extremely large number of edge detection operators available, each designed to be sensitive to certain types of edges. The variables involved in the selection of an edge detection operator include: Edge orientation: (i) The geometry of the operator determines a characteristic direction in which it is most sensitive to edges. Operators can be optimized to look for horizontal, vertical, or diagonal edges; (ii) Noise environment: Edge detection is difficult in noisy images, since both the noise and the edges contain high-frequency content. Attempts to reduce the noise result in blurred and distorted edges. Operators used on noisy images are typically

S. Fidanova, *Ant Colony Optimization and Applications*, Studies in Computational Intelligence 947, https://doi.org/10.1007/978-3-030-67380-2_8

larger in scope, so they can average enough data to discount localized noisy pixels. This results in a less accurate localization of the detected edges; (iii) Edge structure: Not all edges involve a step change in intensity. Effects such as refraction or poor focus can result in objects with boundaries defined by a gradual change in intensity. The edge detection operator needs to be chosen to be responsive to such a gradual change in those cases. There are many ways to perform edge detection. However, the majority of different approaches may be grouped into two categories [2–5]: Gradient based operators (First order edge detection): The gradient approach detects edges by looking for the maximum or minimum in the first derivative of the image. Laplacian based operators (Second-order edge detection): The Laplacian approach searches for zero crossings in the second derivative of the image to find edges. (i) Gradient based edge operators are based on the use of a first order derivative. Robert's, Prewitt's and Sobel are classified as typical operators in [2, 5]. The Robert's cross operator consists of a pair of 2×2 convolution kernels. One kernel is the other rotated by 90 deg. The Prewitt's and Sobel operators consist of a pair of 3×3 convolution kernels as one kernel is the other rotated by 90 deg. The two kernels of the three operators are used for detecting vertical and horizontal edges. These classical operators are easy to operate but highly sensitive to noise. (ii) Laplacian based operators are based on the second order derivative, in particular, on the Laplacian. These operators mark a pixel as an edge at a position where the second derivative of the image function becomes zero [2, 5]. The Laplacian of Gaussian (LoG) or Marr-Hildreth edge detector uses both Gaussian and Laplacian operator so that the Gaussian operator reduces the noise and the Laplacian operator detects the sharp edges. The Marr-Hildreth operator, however, suffers from two main limitations. It generates responses that do not correspond to edges, so-called false edges, and its localization error may be rather serious at curved edges.

In the field of edge detection, the edge detector of Canny is of particular importance [6]. The method was developed by John F. Canny in 1986. It is an optimal edge detection technique as it provides good detection, clear response and good localization. The process of the Canny edge detection algorithm can be broken down into 5 different steps: 1. Apply Gaussian filter to smooth the image in order to remove the noise; 2. Find the intensity gradients of the image; 3. Apply non-maximum suppression as an edge thinning technique; 4. Apply double threshold to determine potential edges; 5. Track edges by hysteresis: Finalize the detection of edges by suppressing all other edges that are weak and not connected to strong edges. Some disadvantages of the edge detection algorithm of Canny and some means to overcome them are presented in [7].

In this chapter we apply multiagent system for image edge detection. The idea comes from Ant Colony Optimization (ACO) methods. ACO approach is applied to solve hard combinatorial optimization problems, which need huge amount of computational resources. We use the swarm intelligence of the ant to look for brightness change and to detect edges. We simulate ant behavior, in a similar way as in optimization applications.

8.2 Image Edge Detection Algorithm Inspired by ACO

The problem of Image edges detection is a problem to find a pixels of the image which correspond to edges. The two dimensional image is represented as $N \times M$ matrix. The element (i, j) of the matrix corresponds to the pixel (i, j), and its value is the pixel intensity. The graph of the problem is as follows. The nodes of the graph corresponds to the pixels. The arcs of the graph connect adjacent nodes. Every internal node has 8 adjacent nodes: up and down nodes, left and right nodes and the four diagonal nodes. If the node (i, j) is internal node, the set of its adjacent nodes is $\{(i-1, j), (i-1, j-1), (i, j-1), (i+1, j-1), (i+1, j), (i+1, j+1), (i, j+1), (i-1, j+1)\}$.

The ants start to create the solution starting from random node. When the ACO algorithm solve optimization problem, every ant create its own solution and at the end of the iteration we compare them and choose the best one for the current best solution. On the next iteration the ants try to find better solutions. When we apply ant idea on image edge detection, the ants try to construct part of the image edges and merging the edges detected by individual ant we construct the image edges. In traditional ACO algorithm for optimization problems, ants construct new solutions taking in to account the regions with good solutions according their experience, marked by the pheromone. In our application in every iteration the ants include new edges in current solution, thus they continue to rebuild the image edges.

In our algorithm an ant starts, to look for edges, from random internal node (i, j), or if the image matrix is $N \times M$, $0 < i < N-1$ and $0 < j < M-1$. The ant move to one of the adjacent nodes according transition probability rule:

$$p_{i,j} = \tau_{ij} * \eta_{ij} \tag{8.1}$$

Where $\eta_{ij} = \frac{V_{ij}}{Vmax}$, $Vmax$ is the maximal value of V_{ij} for $0 < i < N-1$ and $0 < j < M-1$.

$$\begin{aligned} V_{ij} = {} & |I_{i-1,j-1} - I_{i+1,j+1}| + |I_{i-1,j} - I_{i+1,j}| + \\ & |I_{i-1,j+1} - I_{i+1,j-1}| + |I_{i,j-1}, I_{i,j+1}| \end{aligned} \tag{8.2}$$

An ant stops to add new pixels in the solution when $p_{ij} < \varepsilon_1$, where ε_1 is a parameter. When all ants finish to create their edge we delete false edges, new edges consisting only one node. At the end of every iteration, the pheromone of the visited pixels is updated according the rule:

$$\tau_{ij} = \rho * \tau_{ij} + (1-\rho) * (\tau_{init} + \varepsilon) \tag{8.3}$$

where τ_{init} is the initial pheromone value and ε is a small positive number close to 0.

At the next iterations ants again start from random nodes and add new edges on current solution. The algorithm continue till no new edges are detected. The end condition is:

$$|T(k-1) - T(k)| < \varepsilon \tag{8.4}$$

where ε is a parameter and $T(k)$ is the sum of the pheromone of all pixels. If the pheromone of the image is not changed, it means that new edge is not detected.

8.3 Experimental Results

In this section we show some experimental results. The experiments are on well know image called Lena and which is used in image processing as a good example, Fig. 8.1. The image has a size of 256×256 pixels. By the parameter ε_1 we control how detailed to be the edges. We can need different level of details for different use of edge detection. On Fig. 8.2 the ε_1 parameter is equal to 0.05 and to 0.25. We run the algorithm on computer with Pentium processor 2.8 GHz. We find the reported image edges for 4sek. When the value of parameter ε_1 is small the image edges are much more detailed than, when the value of the ε_1 is higher. We observe that our algorithm can find the image edges with big details.

The parameters of our algorithm are:

- $\tau_{init} = 0.5$ – initial pheromone
- $\rho = 0.5$ – evaporation parameter
- $A = 50$ – number of ants
- $\varepsilon = 0.00001$
- $n = 1000$ – number of iterations

The number of iterations is set to be 1000, but because the end condition, the algorithm performs up to 300 iterations (Figs. 8.3, 8.4, 8.5, and 8.6).

We compare our algorithm on other well known test images too as FRUIT-Copy, Fruits, Mandrill and Peppers. We achieved very detailed image edges without beforehand smoothing the image.

Let us compare the image edges achieved by our algorithm with this achieved by ant algorithms, proposed by other authors. Our algorithm finds much more detailed edges comparing with ant based algorithms proposed in [8–10]. The image edges detected by the ant algorithm proposed in [11] are detailed as the image edges detected by our algorithm, but their algorithm detects some false edges too. Other algorithms for images edge detection are wavelet based algorithms [12]. They achieves similar or worst edges detection, but the algorithm complexity is higher. Thees algorithms need smoothing of the original image to find good result. The complexity of the wavelet based algorithms is $O(L \times M \times N)$, where M and N is the image size and L is the length of the used filters. In ACO algorithm we fixed the number of ants to

Fig. 8.1 Lena

Fig. 8.2 Detected edges when $\varepsilon_1 = 0.05$ and $\varepsilon_1 = 0.25$

Fig. 8.3 FRUIT-Copy

Fig. 8.4 Fruits

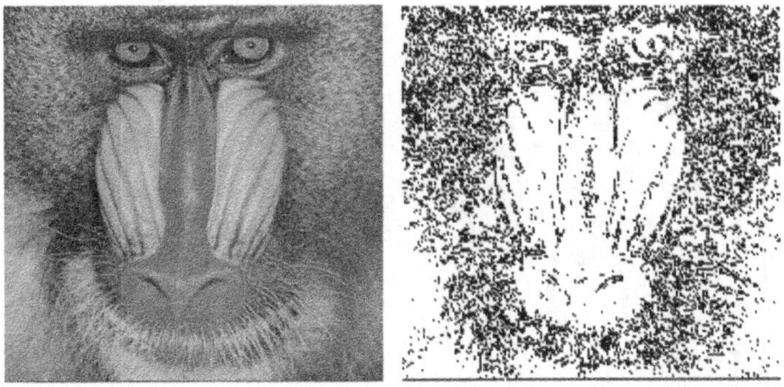

Fig. 8.5 Mandrill

Fig. 8.6 Peppers

be 50 and every one of the ants starts from random pixel and adds new pixels, in the edge set. Thus our algorithm do not need to perambulate all pixels. The complexity of the proposed algorithm is $A \times n \times L_1$, where A is the number of used ants, n is the number of iterations and L_1 is the number of added pixels by one ant. Thus the algorithm complexity is $O(N + M)$.

8.4 Conclusion

In this chapter is presented an ACO based algorithm for image edge detection. By parameters we can control how detailed to be the edges. The algorithm deletes the false edges. Experimental results show the feasibilities of achieved results. Comparison with ACO algorithms proposed by other authors shows that our algorithm find more detailed image edges and less false edges. Comparison with wevlet algorithms shows that our has less complexity. We can conclude that achieved results are very encouraging. Our algorithm is fast and achieves good solutions.

References

1. Fidanova, S., Ilcheva, Z.: Application of ants ideas on image edge detection. Large Scale Scintific Comput., Lect. Notes Comput. Sci. **9374**, 200–207. Springer (2016)
2. Gonzalez, R.C., Woods, R.E.: Digital Image Processing. Prentice-Hall, Inc. (2002)
3. Pratt, W.K.: Digital Image Processing, 2nd edn. John Wiley & Sons (1991)
4. Jain, A.K.: Fundamentals of Digital Image Processing. Prentice-Hall, Inc. (1989)
5. Mlsna, P.A., Rodriguez, J.J.: Gradient and Laplacian-type edge detection. In: Al Bovik (ed.) Handbook of Image and Video Processing, pp. 415–431, Academic Press (2000)
6. Canny, J.F.: A Computational approach to edge detection. IEEE Trans. Pattern Anal. Machine Intell. PAMI-**8**(6), 679–697 (1986)
7. Zhou, P., Ye, W.Q., Wang, Q.: An improved Canny algorithm for edge detection. J. Comput. Inf. Syst. **7**(5), 1516–1523 (2011)
8. Baterina, A.V., Oppus, C.: Image edge detection using ant colony optimization. WSEAS Trans. Signal Process. **6**(8), 58–67 (2010)
9. Tian, J., Yu, W., Xie, S.: An Ant Colony Optimization Algorithm for Image Edge Detection, pp. 751–756. IEEE Congress on Evolutionary Computation, Hong Kong (2008)
10. Nezamabadi-pour, H., Saryazdi, S., Rashedi, E.: Edge detection using ant algorithm. Soft Comput. **10**(7), 623–628 (2006)
11. Jevtic, A., Li, B.: Ant algorithm for adaptive edge detection. In: Search Algorithms for Engineering Optimization, Chapter 2, INTECH Publisher (2013)
12. Zhang, Z., Ma, S., Liu, H., Gong, Y.: An edge detection approach based on directional wavelet transform. J. Comput. Math. Appl. **57**(8), 1265–1271 (2009)

Chapter 9
ACO for Workforce Planning

The text of this chapter is based on [1, 2]. Planing the workforce is difficult and important decision making industrial problem. It includes several level of complexity: selection and assignment. First the employers need to be selected from large set of available workers. The second is selected employers to be assigned to the jobs to be performed. The aim is to fulfil the work requirements with minimal assignment cost. The human resource management is one of the important parts of the production organization.

It is impossible to apply exact algorithms on realistic instances, because of the complexity of the problem. In [3, 4], deterministic workforce planing problem is studied. In the work [3] workforce planning models that contain non-linear models of human learning are reformulated as mixed integer programs. The authors show that the mixed integer program is much easier to solve than the non-linear program. A model of workforce planning is considered in [4]. The model includes workers differences, as well as the possibility of workers training and upgrading. A variant of the problem with random demands is proposed in [5, 6]. In [5] a two-stage stochastic program for scheduling and allocating cross-trained workers is proposed considering a multi-department service environment with random demands. In to some problems uncertainty has been employed [7–11]. In this case the corresponding objective function and given constraints is converted into crisp equivalents and then the model is solved by traditional methods [9] or the considered uncertain model is transformed into an equivalent deterministic form as it is shown in [10]. Most of the authors simplifies the problem by omitting some of the constraints. Some conventional methods can be applied on workforce planning problem as mixed linear programming [12], decomposition method [6]. However, for the more complex non-linear workforce planning problems, the convex methods are not applicable. There are some applications of heuristic methods including genetic algorithms [13, 14], memetic algorithms [15], scatter search [13], etc. Metaheuristic algorithms usually are applied when the problem is complex and has a very strong constraints [16, 17]. In this work we propose an Ant Colony Optimization (ACO) algorithm for workforce planning problem

S. Fidanova, *Ant Colony Optimization and Applications*, Studies in Computational Intelligence 947, https://doi.org/10.1007/978-3-030-67380-2_9

[1]. So far the ACO algorithm is proved to be very effecting solving various complex optimization problems [18, 19]. We focus on optimization of the algorithm parameters and finding the minimal number of ants which is need to find the best solution. It is known that when the number of ant is doubles, the computational time and the used memory are doubles. When the number of iteration is doubles, only the computational time is doubles. We look for a minimal product between number of ants and number of iterations that is needed to find the best solution.

9.1 Problem Formulation

On this chapter we use the description of workforce planing problem given by Glover et al. [20]. There is a set of jobs $J = \{1, \dots, m\}$, which must be completed during a fixed period. Each job j requires d_j hours to be completed. The set of available workers is $I = \{1, \dots, n\}$. Every worker must perform every of assigned to him job minimum h_{min} hours, reason of efficiency. The worker i is available s_i hours. The maximal number of assigned jobs to a same worker is j_{max}. The workers have different skills and the set A_i shows the jobs that the worker i is qualified to perform. The maximal number of workers which can be assigned during the planed period is t or at most t workers may be selected from the set I of workers and the selected workers can be capable to complete all the jobs. The aim is to find feasible solution that minimizes the assignment cost.

Every worker i and job j are related with cost c_{ij} of assigning the worker to the job. The workforce planing problem can be represented mathematicaly as follows:

$$x_{ij} = \begin{cases} 1 \text{ if the worker } i \text{ is assigned to job } j \\ 0 \text{ otherwise} \end{cases}$$

$$y_i = \begin{cases} 1 \text{ if worker } i \text{ is selected} \\ 0 \text{ otherwise} \end{cases}$$

$$z_{ij} = \text{number of hours that worker } i$$

$$\text{is assigned to perform job } j$$

$$Q_j = \text{set of workers qualified to perform job } j$$

$$\text{Minimize} \sum_{i \in I} \sum_{j \in A_i} c_{ij}.x_{ij} \tag{9.1}$$

Subject to

$$\sum_{j \in A_i} z_{ij} \leq s_i.y_i \quad i \in I \tag{9.2}$$

$$\sum_{i \in Q_j} z_{ij} \geq d_j \quad j \in J \tag{9.3}$$

$$\sum_{j \in A_i} x_{ij} \leq j_{max}.y_j \quad i \in I \tag{9.4}$$

$$h_{min}.x_{ij} \leq z_{ij} \leq s_i.x_{ij} \quad i \in I, j \in A_i \tag{9.5}$$

$$\sum_{i \in I} y_i \leq t \tag{9.6}$$

$$\begin{aligned} &x_{ij} \in \{0, 1\} \; i \in I, j \in A_i \\ &y_i \in \{0, 1\} \;\; i \in I \\ &z_{ij} \geq 0 \qquad i \in I, j \in A_i \end{aligned}$$

The assignment cost is the objective function of this problem. The number of hours for each selected worker is limited (inequality 9.2). The work must be done in full (inequality 9.3). The number of the jobs, that every worker can perform is limited (inequality 9.4). There is minimal number of hours that every job must be performed by every assigned worker to can work efficiently (inequality 9.5). The number of assigned workers is limited (inequality 9.6).

Different objective functions can be optimized with the same model. In this paper our aim is to minimize the total assignment cost. If $\tilde{c}_{ij}$ is the cost the worker i to performs the job j for one hour, than the objective function can minimize the cost of the hall jobs to be finished (on hour basis).

$$f(x) = \text{Min} \sum_{i \in I} \sum_{j \in A_i} \tilde{c}_{ij}.x_{ij}.z_{ij} \tag{9.7}$$

Some worker can have preference to perform part of the jobs he is qualified and the objective function can be to maximize the satisfaction of the workers preferences or to maximize the minimum preference value for the set of selected workers.

As we mentioned above in this paper the assignment cost is minimized (Eq. 9.1). The workforce planning problem is difficult to be solved because of very restrictive constraints especially the relation between the parameters h_{min} and d_j. When the problem is structured (d_j is a multiple of h_{min}), it is more easier to find feasible solution, than for unstructured problems (d_j and h_{min} are not related).

9.2 ACO Algorithm for Workforce Planning

One of the essential point of the ant algorithm is the proper representation of the problem by graph. In our case the graph of the problem is 3 dimensional and the node (i, j, z) denotes to worker i to be assigned to the job j for time z. At the beginning of every iteration each ant starts to construct its solution, from random node of the graph of the problem. For each ant are generated three random numbers. The first random number is in the interval $[0, \ldots, n]$ and corresponds to the worker we assign. The second random number is in the interval $[0, \ldots, m]$ and corresponds to the job which this worker will perform. The third random number is in the interval $[h_{min}, \ldots, \min\{d_j, s_i\}]$ and corresponds to the number of hours worker i is assigned to perform the job j. Subsequently, the ant applies the transition probability rule to include next nodes in the partial solution, until the solution is completed.

We propose the following heuristic information:

$$\eta_{ijl} = \begin{cases} l/c_{ij} & l = z_{ij} \\ 0 & otherwise \end{cases} \tag{9.8}$$

This heuristic information stimulates to assign the most cheapest worker as longer as possible. The ant chooses the node with the highest probability. When an ant has several possibilities for next node (several candidates have the same probability to be chosen), the next node is chosen randomly among them.

When a new node is included we take in to account how many workers are assigned currently, how many time slots every worker is currently assigned and how many time slots are currently assigned per job. When some move of the ant does not meet the problem constraints, the probability of this move is set to 0. If it is impossible to include new nodes from the graph of the problem (for all nodes the value of the transition probability is 0), the construction of the solution stops. When the constructed solution is feasible the value of the objective function is the sum of the assignment cost of the assigned workers. If the constructed solution is not feasible, the value of the objective function is set to -1.

Only the ants, which constructed feasible solution are allowed to add new pheromone to the elements of their solutions. The newly added pheromone is equal to the reciprocal value of the objective function:

$$\Delta\tau_{i,j} = \frac{\rho - 1}{f(x)} \tag{9.9}$$

Thus, the nodes of the problem graph, which belong to better solutions (less value of the objective function) receive more pheromone than the other nodes and become more desirable in the next iteration.

At the end of every iteration we compare the best solution with the best so far solution. If the best solution from the current iteration is better than the best so far solution (global best solution), we update the global best solution with the current iteration best solution.

The end condition used in our ACO algorithm is the number of iterations.

The software, which realizes the algorithm is written in C and is run on Pentium desktop computer 2.8 GHz with 4 GB of memory.

We use the artificially generated problem instances considered in [13]. The test instances characteristics are shown in Table 9.1.

The set of test problems consists of ten structured and ten unstructured problems. The problem is structured when d_j is proportional to h_{min}. The structured problems are enumerated from $S01$ to $S10$ and unstructured problems are enumerated from $U01$ to $U10$.

The number of iterations (stopping criteria) is fixed to be 100. The parameter settings of our ACO algorithm are shown in Table 9.2. The values are fixed experimentally.

The algorithm is stochastic and from a statistical point of view it needs to be run minimum 30 times to guarantee the robustness of the average results. We perform 30 independent runs of the algorithm. Afterwards statistical analysis of the results applying ANOVA test was done.The test shows that there is significant difference between the results achieved by different methods, or the results are not statistically the same.

Let us compare the numerical results achieved by our ACO algorithm and those achieved by genetic algorithm (GA) and scatter search (SS) presented in [13]. Table 9.3 shows the achieved results for structured instances, while Table 9.4 shows the obtained results for unstructured instances.

We observe that ACO algorithm outperforms the other two algorithms. ACO is a constructive method and when the graph of the problem and heuristic information are appropriate and they represent the problem in a good way, it can help a lot of for better

Table 9.1 Test instances characteristics

Parameters	Value
n	20
m	20
t	10
s_i	[50,70]
j_{max}	[3,5]
h_{min}	[10,15]

Table 9.2 ACO parameter settings

Parameters	Value
Number of iterations	100
ρ	0.5
τ_0	0.5
Number of ants	20
a	1
b	1

Table 9.3 Average results for structured problems

Test problem	Objective function value		
	SS	GA	ACO
$S01$	936	963	807
$S02$	952	994	818
$S03$	1095	1152	882
$S04$	1043	1201	849
$S05$	1099	1098	940
$S06$	1076	1193	869
$S07$	987	1086	812
$S08$	1293	1287	872
$S09$	1086	1107	793
$S10$	945	1086	825

Table 9.4 Average results for unstructured problems

Test problem	Objective function value		
	SS	GA	ACO
$U01$	1586	1631	814
$U02$	1276	1264	845
$U03$	1502	1539	906
$U04$	1653	1603	869
$U05$	1287	1356	851
$U06$	1193	1205	873
$U07$	1328	1301	828
$U08$	1141	1106	801
$U09$	1055	1173	768
$U10$	1178	1214	818

algorithm performance and achieving good solutions. Our graph of the problem has a star shape. Each worker and jobs are linked with several nodes, corresponding to the time, for which the worker is assigned to perform this job. The proposed heuristic information stimulates the cheapest workers to be assigned for longer time. It is a greedy strategy. After the first iteration the pheromone level reflects the experience of the ants during the searching process thus affects the strategy. The elements of good solutions accumulate more pheromone, during the algorithm performance, than others and become more desirable in the next iterations.

Now we will compare the execution time of the proposed ACO algorithm with the execution time of the other two algorithms—GA and SS [13]. The algorithms are run on similar computers. In Table 9.5 the parameters of the GA and SS algorithms are presented.

Table 9.5 Algorithms parameter settings, as given in [13]

Parameters	Genetic algorithm
Population size	400
Crossover rate	0.8
Mutation rate	0.2
Parameters	Scatter search
Initial population	15
Reference set update and creation	8
Subset generated	All 2-elements subsets
ρ_i	0.1

Table 9.6 Average time for structured problems

Test problem	Execution time, *s*		
	SS	GA	ACO
$S01$	72	61	26
$S02$	49	32	21
$S03$	114	111	22
$S04$	86	87	25
$S05$	43	40	21
$S06$	121	110	23
$S07$	52	49	23
$S08$	46	42	24
$S09$	70	67	20
$S10$	105	102	22

The average execution times over 30 runs of each of the algorithms are reported in Tables 9.6 and 9.7.

It is seen that the ACO algorithm finds the solution faster than GA and SS. Considering the execution time the GA and SS algorithms have similar performance. By the numerical results presented in Tables 9.3, 9.4, 9.6 and 9.7 we can conclude that ACO algorithm gives very encouraging results. It achieves better solutions in shorter time than the other two algorithms, SS and GA. If we compare the used memory, the ACO algorithm uses less memory than GA (GA population size is 400 individuals) and similar memory to SS (initial population size is 15 and reference set is 8 individuals) [13].

Table 9.7 Average time for unstructured problems

Test problem	Execution time, *s*		
	SS	GA	ACO
*U*01	102	95	22
*U*02	94	87	20
*U*03	58	51	20
*U*04	83	79	20
*U*05	62	57	23
*U*06	111	75	22
*U*07	80	79	21
*U*08	123	89	20
*U*09	75	72	26
*U*10	99	95	20

9.3 Hybrid ACO Algorithm for Workforce Planning

Our main contribution in this section is the hybridization of the ACO algorithm with a local search procedure. The aim of the local search is to decrease the time to find the best solution and eventually to improve the achieved solutions.

We apply local search procedure only on infeasible solutions and only one time disregarding the new solution is feasible or not. Thus, our local search is not time consuming. The workforce planning is a problem with strong constraints and part of the ants do not succeed to find feasible solution. With our local search procedure the possibility to find feasible solution increases and thus increases the chance to improve current solution.

If the solution is not feasible we remove part of the assigned workers and after that we assign in their place new workers. The workers which will be removed are chosen randomly. On this partial solution we assign new workers applying the rules of ant algorithm. The ACO algorithm is a stochastic algorithm, therefore the new constructed solution is different from previous one with a high probability.

In this section test results are reported and comparison between ACO algorithm with local search and ACO algorithm without local search procedure is done. The software which realizes the algorithm is written in C computer language and is run on Pentium desktop computer 2.8 GHz with 4 GB RAM.

We use the artificially generated problem instances considered in [13], like in the previous section. The number of iterations is a stopping criteria for our hybrid ACO algorithm. The number of iterations is fixed to be maximum 100.

The workforce problem has very restrictive constraints. Therefore only 2–3 of the ants, per iteration, find feasible solution. Sometimes exist iterations without any feasible solution. Its complicates the search process. Our aim is to decrease the number of unfeasible solutions and thus to increase the possibility ants to find good solutions and so to decrease needed number of iterations to find good solution. Our local search procedure is not time consuming, because we apply it only on unfeasible solutions and only one time, disregarding if the ant find feasible or unfeasible solution after it. We observe that after the local search procedure applied on the first iteration, the number of unfeasible solutions in a next iterations decrease. It is another reason the calculation time does not increase significantly. We are dealing with four cases: without local search procedure (ACO); local search procedure when the number of removed workers is quarter from the number of all assigned workers (ACO quarter); local search procedure when the number of removed workers is half from the number of all assigned workers (ACO half); local search procedure when all assigned workers are removed and the solution is constructed from the beginning (ACO restart).

We perform 30 independent runs with every one of the four cases, because the algorithm is stochastic and to guarantee the robustness of the average results. We apply ANOVA test for statistical analysis to guarantee the significance of the achieved results.

We are interested of the number of iterations for finding the best result. It can be very different for different test problems, so we will use ranking of the algorithms. The variant of our hybrid algorithm is on the first place, if it achieves the best solution with less average number of iterations over 30 runs, according other cases and we assign to it 1, we assign 2 to the case on the second place, 3 to the case on the third place and 4 to the case with most number of iterations. On some cases can be assigned same numbers if the number of iterations to find the best solution is the same. We sum the ranking of the cases over all 20 test problems to find final ranking of the different cases of the hybrid algorithm.

We observe that the local search procedure decreases the number of unfeasible solutions, found by traditional ACO algorithm in the next iterations, thus when the number of iterations increase, the need of local search procedure decreases. So this local search procedure is not time consuming. On Table 9.8 we report the achieved ranking of different cases of our hybrid algorithm. As we mentioned above, with ACO quarter we call the case when quarter of the workers are removed. ACO half is the case when half of the workers are removed. ACO restart is the case when all workers are removed. It is like to restart the solution construction, to construct the solution from the beginning.

One of the main questions is how many worker to remove from unfeasible solution, so that the new constructed solution to be feasible and close to the best one. We

Table 9.8 Hybrid ACO ranking

	ACO	ACO quarter	ACO half	ACO restart
First place	4 times	4 times	8 times	8 times
Second place	4 times	4 times	7 times	6 times
Third place	8 times	6 times	4 times	3 times
Forth place	4 times	6 times	1 times	3 times
Ranking	52	54	38	41

calculate the ranking, regarding the average number of iterations to find best solution over 30 runs of the test. When more than half of the ants find unfeasible solutions, the deviation from the average is larger compared to the tests when the most of the ants achieve feasible solutions. Table 9.8 shows that the local search procedure decreases the number of iterations needed to find the best solution, when more than half of the workers are removed. The traditional ACO algorithm and hybrid ACO with removing quarter of the workers are 4 times on the first place when either, by chance, the algorithm find the best solution on the first iteration, or all ants find feasible solutions. We observe that the both cases are on the third and forth place 12 times. This means that removing less than half of the workers is not enough to construct feasible solution. The ACO algorithm with removed half of the workers 15 times is on the first or second place and only one time is on the fourth place, which means that it performs much better than previous two cases. When all workers are removed the achieved ranking is similar to the case when half of the workers are removed. Let the maximal number of assigned workers is t. Thus the every one of the solutions consists about t workers. If all of the workers are removed, the ant need to add new workers on their place which number is about t. When half of the workers are removed, then the ant will add about $t/2$ new workers. The calculation time to remove and add about $t/2$ workers is about two times less than to remove and add about t workers. Thus we can conclude that the local search procedure with removing half of the workers performs better than other cases.

Another way for comparison is the calculation time. For every test problem and every case we calculate the average time to achieve best solution over 30 runs. In Table IV we did similar ranking as in Table 9.8, but taking in to account the calculation time instead number of iterations. After we sum the average time for finding the best solution for every case over all test problems. Regarding Table 9.9 the ranking according the time is similar to the ranking according to the number of iterations from Table 9.8. The best performance is when half of the worker are removed in the local search procedure and the worst performance is when quarter of the workers are removed. The local search procedure with removing half of the worker is on the first place 10 times and on the forth place only 2 times. The local search procedure with removing quarter of the workers is on the first place 3 times and on the forth place 8 times. Regarding the calculation time the local search procedure with removing half of the workers again is the best, but the worst is the local search procedure

Table 9.9 Hybrid ACO comparison according calculation time

	ACO	ACO quarter	ACO half	ACO restart
First place	4 times	3 times	10 times	6 times
Second place	7 times	5 times	5 times	5 times
Third place	8 times	4 times	3 times	4 times
Forth place	1 times	8 times	2 times	5 times
Average time	82.244 s	93.98 s	79.63 s	103.012 s

with removing all workers. Reconstructing a solution from the beginning takes more time than to reconstruct partial solution, therefore ACO algorithm with local search procedure removing all workers performs worst. The results from Table 9.9 show that removing only quarter of the workers from the solution is not enough for construction of good solution and is time consuming comparing with traditional ACO algorithm. According the both types of comparison, ranking and calculation time ACO with local search procedure removing half of the assigned workers performs best.

9.4 Influence of ACO Parameters on Algorithm Performance

In this section we analyze the ACO performance according to the number of ants and the quality of the achieved solutions. We use the same artificially generated problem instances, considered in [13].

If the number of ants of ACO algorithm increases, the computational time and the used memory increase proportionally. If the number of iterations increases, only the computational time increases. If the computational time is fixed and we vary only the number of ants it means that we vary the number of iteration too, but in opposite direction, or if the time is fixed it is equivalent to fixing the product of number of ants and number of iterations.

We apply number of ants from the set {5, 10, 20, 40} and respectively, number of iteration—{400, 200, 100, 50}. Because of stochastic nature of the algorithm we run the ACO algorithm for all 20 test problems with each one of the four ACO algorithms ($ACO_{5\times 400}$, $ACO_{10\times 200}$, $ACO_{20\times 100}$ and $ACO_{40\times 50}$) 30 times. We look for the maximal number of iterations, within the fixed computational time, which is needed to find the best solution. We compare the product between the number of ants and number of required iterations for the best performed ACO algorithm. The parameter settings for our ACO algorithm are shown in Table 9.10.

Tables 9.11 and 9.12 show the resulting product between the number of ants and number of iterations that have been used to find the best solution. When the observed product is the same for different ACO algorithms (with different number of ants) the

Table 9.10 ACO parameter settings

Parameters	Value
Number of iterations	400, 200, 100, 50
ρ	0.5
τ_0	0.5
Number of ants	5, 10, 20, 40
a	1
b	1

Table 9.11 Computational results for structured problems

Test problem	Product ants × iterations			
	$ACO_{5\times400}$	$ACO_{10\times200}$	$ACO_{20\times100}$	$ACO_{40\times50}$
$S01$	195	200	260	**120**
$S02$	**195**	330	340	640
$S03$	**475**	490	580	1160
$S04$	**1540**	1540	1540	1560
$S05$	415	**320**	420	920
$S06$	**165**	250	420	520
$S07$	**570**	720	860	880
$S08$	**1125**	1130	1140	1120
$S09$	**855**	860	720	880
$S10$	**230**	230	520	840

Table 9.12 Computational results for unstructured problems

Test problem	Product ants×iterations			
	$ACO_{5\times400}$	$ACO_{10\times200}$	$ACO_{20\times100}$	$ACO_{40\times50}$
$U01$	**330**	340	340	440
$U02$	**160**	340	340	400
$U03$	1000	1000	**560**	640
$U04$	**760**	760	820	820
$U05$	295	295	**280**	280
$U06$	945	940	**720**	920
$U07$	**550**	550	580	760
$U08$	160	300	220	**120**
$U09$	**570**	950	920	840
$U10$	**485**	550	600	1120

best results is this produced by ACO with lesser number of ants, because in this case the used memory is less. The best results are shown in bold.

Let us discuss the results reported in Tables 9.11 and 9.12. Regarding the structured problems, for 8 of them the best execution time is when the number of ants is 5. Only for two of the test problems ($S01$ and $S05$) the execution time is better for 40 and 10 ants, respectively, but the result is close to this achieved with 5 ants. Regarding unstructured test problems, for 6 of them the best execution time is achieved when the number of used ants is 5. For three test problems ($U03$, $U05$ and $U06$) the best results are achieved using ACO with 20 ants. For test $U05$, the result is quite close to the result with 5 ants. Only for test $U03$ the difference with the results using 5 ants is significant. Thus, we can conclude that for this problem the best algorithm performance, using less computational resources, is when the number of used ants is 5.

9.5 InterCriteria Analysis

In this section we use ICrA to obtain some additional knowledge about considered four ACO algorithms. Based on the results in Tables 9.11 and 9.12 we construct the following index matrix, Eq. (9.10):

	$ACO_{5\times400}$	$ACO_{10\times200}$	$ACO_{20\times100}$	$ACO_{40\times50}$
$S01$	195	200	260	120
$S02$	195	330	340	640
$S03$	475	490	580	1160
$S04$	1540	1540	1540	1560
$S05$	415	320	420	920
$S06$	165	250	420	520
$S07$	570	720	860	880
$S08$	1125	1130	1140	1120
$S09$	855	860	720	880
$S10$	230	230	520	840
$U00$	775	780	780	2060
$U01$	330	340	340	440
$U02$	160	340	340	400
$U03$	1000	1000	560	640
$U04$	760	760	820	820
$U05$	295	295	280	280
$U06$	945	940	720	920
$U07$	550	550	580	760
$U08$	160	300	220	120
$U09$	570	950	920	840
$U10$	485	550	600	1120

(9.10)

From Eq. (9.10) it can be seen that the ICrA objects (S01, S02, ..., S10, U00, ..., U05) are the different test problems and the ICrA criteria ($ACO_{5\times400}$, $ACO_{10\times200}$, $ACO_{20\times100}$ and $ACO_{40\times50}$) are the ACO algorithms with different number of ants.

After application of ICrA, using the software ICrAData [21], to index matrix Eq. (9.10) we obtained the two index matrices with the relations between considered four criteria. The resulting index matrices for $\mu_{C,C'}$, $\nu_{C,C'}$ and $\pi_{C,C'}$ values are shown in Eqs. (9.11), (9.12) and (9.13).

$\mu_{C,C'}$	$ACO_{5\times400}$	$ACO_{10\times200}$	$ACO_{20\times100}$	$ACO_{40\times50}$
$ACO_{5\times400}$	1	0.88	0.78	0.74
$ACO_{10\times200}$	0.88	1	0.78	0.71
$ACO_{20\times100}$	0.78	0.78	1	0.81
$ACO_{40\times50}$	0.74	0.71	0.81	1

(9.11)

$\nu_{C.C'}$	$ACO_{5\times400}$	$ACO_{10\times200}$	$ACO_{20\times100}$	$ACO_{40\times50}$
$ACO_{5\times400}$	0	0.10	0.18	0.23
$ACO_{10\times200}$	0.10	0	0.20	0.27
$ACO_{20\times100}$	0.18	0.20	0	0.14
$ACO_{40\times50}$	0.23	0.27	0.14	0

(9.12)

$\pi_{C.C'}$	$ACO_{5\times400}$	$ACO_{10\times200}$	$ACO_{20\times100}$	$ACO_{40\times50}$
$ACO_{5\times400}$	0	0.02	0.04	0.03
$ACO_{10\times200}$	0.02	0	0.03	0.03
$ACO_{20\times100}$	0.04	0.03	0	0.05
$ACO_{40\times50}$	0.03	0.03	0.05	0

(9.13)

The obtained ICrA results are visualized on Fig. 9.1 within the specific triangular geometrical interpretation of IFSs.

The results show that the following criteria pairs, according to [22], are in:

- positive consonance:
  $ACO_{5\times400}$-$ACO_{10\times200}$—with degree of "agreement"
  $\mu_{C,C'} = 0.88$;
- weak positive consonance:
  $ACO_{20\times100}$-$ACO_{40\times50}$,
  $ACO_{20\times100}$-$ACO_{5\times400}$ and
  $ACO_{20\times100}$-$ACO_{10\times200}$—with degree of "agreement"
  $\mu_{C,C'} = 0.81$, $\mu_{C,C'} = 0.78$ and $\mu_{C,C'} = 0.78$, respectively;

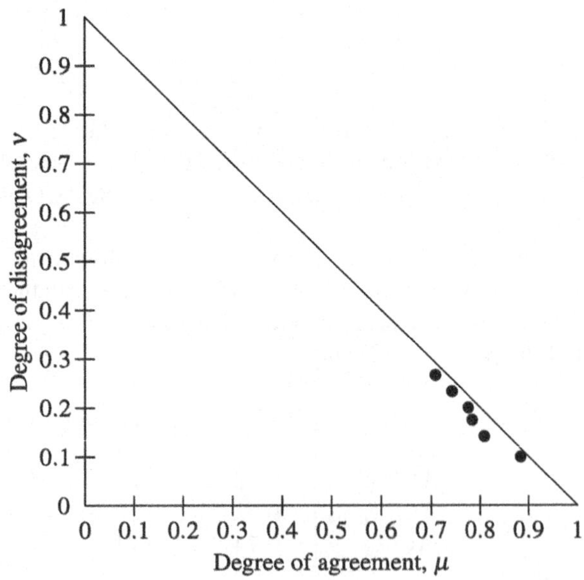

Fig. 9.1 Presentation of ICrA results in the intuitionistic fuzzy interpretation triangle

- weak dissonance:
  $ACO_{40\times 50}$-$ACO_{5\times 400}$ and
  $ACO_{40\times 50}$-$ACO_{10\times 200}$—with degree of "agreement"
  $\mu_{C,C'} = 0.74$ and $\mu_{C,C'} = 0.71$, respectively.

ACO algorithms with close values of number of ants (5 and 10, 10 and 20, 20 and 40) show similar performance. The same result is obtained for ACO algorithms with 5 and 20 ants. The ACO algorithms with bigger difference in the number of ants (5 and 40, 10 and 40) are in weak dissonance, i.e. their performance is not similar (Tables 9.13 and 9.14).

In [23] the author propose to rank the criteria pairs in both dimensions simultaneously (degrees of "agreement" $\mu_{C,C'}$ and "disagreement" $\nu_{C,C'}$ of the intuitionistic fuzzy pairs). This can be done by calculation for each point in the Fig. 9.1 its distance from the point $\langle 1, 0\rangle$. The formula for the distance $d_{C,C'}$ of the pair C, C' to the $\langle 1, 0\rangle$ point is:

$$d_{C,C'} = \sqrt{(1 - \mu_{C,C'})^2 + \nu_{C,C'}{}^2} \tag{9.14}$$

The results are presented in Table 9.14.

Table 9.13 Consonance and dissonance scale, according to [22]

Interval of $\mu_{C,C'}$	Meaning
[0.00–0.05]	Strong negative consonance
(0.05–0.15]	Negative consonance
(0.15–0.25]	Weak negative consonance
(0.25–0.33]	Weak dissonance
(0.33–0.43]	Dissonance
(0.43–0.57]	Strong dissonance
(0.57–0.67]	Dissonance
(0.67–0.75]	Weak dissonance
(0.75–0.85]	Weak positive consonance
(0.85–0.95]	Positive consonance
(0.95–1.00]	Strong positive consonance

Table 9.14 Index matrix of criteria distances from point $\langle 1, 0\rangle$

	$ACO_{5\times 400}$	$ACO_{10\times 200}$	$ACO_{20\times 100}$	$ACO_{40\times 50}$
$ACO_{5\times 400}$	0	0.154	0.279	0.348
$ACO_{10\times 200}$	0.154	0	0.301	0.395
$ACO_{20\times 100}$	0.279	0.301	0	0.238
$ACO_{40\times 50}$	0.348	0.395	0.238	0

In this case the criteria pairs (different ACO algorithms) are ordered according to their $d_{C,C'}$ sorted in decreasing order, as follows:

- $ACO_{5\times400}$–$ACO_{10\times200}$;
- $ACO_{20\times100}$–$ACO_{40\times50}$;
- $ACO_{5\times400}$–$ACO_{20\times100}$;
- $ACO_{10\times200}$–$ACO_{20\times100}$;
- $ACO_{5\times400}$–$ACO_{40\times50}$;
- $ACO_{40\times50}$–$ACO_{10\times200}$.

As it can be seen the similar performance of ACO algorithms with 5 and 10, and 20 and 40 ants is approved by these results too. The similarity between ACO with 5 and ACO with 20 ants is observed again. The next three criteria pairs, are ranked in the same manner, too. Thus the obtained ICrA results are confirmed by two different approaches—when using the scale, proposed in [22] and according simultaneously to the degrees of "agreement" $\mu_{C,C'}$ and "disagreement" $\nu_{C,C'}$ of the intuitionistic fuzzy pairs [23].

On the other hand, ICrA confirms the conclusion that for this problem the best algorithm performance, i.e., using less computational resources, is shown ACO algorithm with five number of ants.

9.6 Conclusion

In this chapter we propose ACO algorithm for solving workforce planning problem. We compare the performance of our algorithm with other two metahuristic methods, genetic algorithm and scatter search. The comparison is done by various criteria. We observed that ACO algorithm achieves better solutions than the two other algorithms. Regarding the execution time the ACO algorithm is faster. The ACO population consists of 20 individuals and the memory used by the algorithm is similar to the one used by the SS and less than the memory used by the GA. We propose appropriate local search procedure, which decrease the running time for achieving the best solution. The influence of the ACO parameters on algorithm performance is learned. At the end InterCriteria analyzes is applied to obtain some additional knowledge about relations of algorithms' parameters.

References

1. Fidanova, S., Luquq, G., Roeva, O., Paprzycki, M., Gepner, P.: Ant colony optimization algorithm for workforce planning. FedCSIS'2017, pp. 415–419. IEEE Xplorer (2017)
2. Fidanova, S., Luquq, G., Roeva, O., Paprzycki, M., Gepner, P.: Hybrid Ant colony optimization algorithm for workforce planning. Ann. Comput. Sci. Inf. Syst. 15, 233–236 (2018). ISSN:2300-5963, https://doi.org/10.15439/2018F47

3. Hewitt, M., Chacosky, A., Grasman, S., Thomas, B.: Integer programming techniques for solving non-linear workforce planning models with learning. Eur. J. Oper. Res. **242**(3), 942–950 (2015)
4. Othman, M., Bhuiyan, N., Gouw, G.: Integrating workers' differences into workforce planning. Comput. Ind. Eng. **63**(4), 1096–1106 (2012)
5. Campbell, G.: A two-stage stochastic program for scheduling and allocating cross-trained workers. J. Oper. Res. Soc. **62**(6), 1038–1047 (2011)
6. Parisio, A., Jones, C.N.: A two-stage stochastic programming approach to employee scheduling in retail outlets with uncertain demand. Omega **53**, 97–103 (2015)
7. Hu, K., Zhang, X., Gen, M., Jo, J.: A new model for single machine scheduling with uncertain processing time. J. Intell. Manuf. **28**(3), 717–725 (2015)
8. Li, R., Liu, G.: An uncertain goal programming model for machine scheduling problem. J. Intell. Manuf. **28**(3), 689–694 (2014)
9. Ning, Y., Liu, J., Yan, L.: Uncertain aggregate production planning. Soft Comput. **17**(4), 617–624 (2013)
10. Yang, G., Tang, W., Zhao, R.: An uncertain workforce planning problem with job satisfaction. Int. J. Mach. Learn. Cybern. Springer. https://doi.org/10.1007/s13042-016-0539-6 (2016)
11. Zhou, C., Tang, W., Zhao, R.: An uncertain search model for recruitment problem with enterprise performance, J. Intell. Manuf. **28**(3), 295–704 (2014). Springer
12. Easton, F.: Service completion estimates for cross-trained workforce schedules under uncertain attendance and demand. Prod. Oper. Manage. **23**(4), 660–675 (2014)
13. Alba, E., Luque, G., Luna, F.: Parallel metaheuristics for workforce planning. J. Math. Model. Algorithms **6**(3), 509–528 (2007). Springer
14. Li, G., Jiang, H., He, T.: A genetic algorithm-based decomposition approach to solve an integrated equipment-workforce-service planning problem. Omega **50**, 1–17 (2015)
15. Soukour, A., Devendeville, L., Lucet, C., Moukrim, A.: A Memetic algorithm for staff scheduling problem in airport security service. Expert Syst. Appl. **40**(18), 7504–7512 (2013)
16. Isah, O.R., Usman, A.D., Tekanyi, A.M.S.: A hybrid model of PSO algorithm and artificial neural network for automatic follicle classification. Int. J. Bioautomation **21**(1), 43–58 (2017)
17. Zeng, J., Li, Y.: The use of adaptive genetic algorithm for detecting Kiwifruits variant subculture seedling. Int. J. Bioautomation **21**(4), 349–356 (2017)
18. Grzybowska, K., Kovács, G.: Sustainable Supply chain – Supporting tools. Proceedings of the 2014 Federated Conference on Computer Science and Information Systems 2, 1321–1329 (2014)
19. Fidanova, S., Roeva, O., Paprzycki, M., Gepner, P.: InterCriteria Analysis of ACO start strategies. Proceedings of the 2016 Federated Conference on Computer Science and Information Systems, 547–550 (2016)
20. Glover, F., Kochenberger, G., Laguna, M., Wubbena, T.: Selection and assignment of a skilled workforce to meet job requirements in a fixed planning period. In: MAEB04, pp. 636–641 (2004)
21. Ikonomov, N., Vassilev, P., Roeva, O.: ICrAData software for InterCriteria analysis. Int. J. Bioautomation **22**(2) (2018)
22. Atanassov, K., Atanassova, V., Gluhchev, G.: InterCriteria analysis: Ideas and problems. Notes on Intuitionistic Fuzzy Sets **21**(1), 81–88 (2015)
23. Atanassova, V.: Interpretation in the intuitionistic fuzzy triangle of the results, obtained by the interCriteria analysis. In: Proceedings of the 9th Conference of the European Society for Fuzzy Logic and Technology (EUSFLAT), pp. 1369–1374 (2015)

The manufacturer's authorised representative in the EU is Springer Nature Customer Service Centre GmbH, Europaplatz 3, 69115 Heidelberg, Germany. If you have any concerns regarding our products, please contact ProductSafety@springernature.com

Printed and bound by CPI Group (UK) Ltd, Croydon, CR0 4YY

30/09/2026

02225307-0006